THE SOCIAL RELATIONS OF PHYSICS, MYSTICISM, AND MATHEMATICS

A PALLAS PAPERBACK

SAL RESTIVO

Science and Technology Studies Division,
Rensselaer Polytechnic Institute, Troy, NY 12181, U.S.A.

THE SOCIAL RELATIONS OF PHYSICS, MYSTICISM, AND MATHEMATICS

Studies in Social Structure, Interests, and Ideas

D. REIDEL PUBLISHING COMPANY

A MEMBER OF THE KLUWER ACADEMIC PUBLISHERS GROUP

DORDRECHT / BOSTON / LANCASTER

Library of Congress Cataloging in Publication Data

CIP

Restivo, Sal P.
The social relations of physics, mysticism, and mathematics.

(Episteme ; v. 10)
Includes bibliographical references and index.
1. Science–Social aspects. 2. Physics–Social aspects.
3. Mathematics–Social aspects. 4. Mysticism–Social aspects.
I. Title. II. Series: Episteme (D. Reidel) ; v. 10.
Q175.5.R47 1983 306'.45 83-3220
ISBN 90-277-1536-X
ISBN 90-277-2084-3 (pbk)

Published by D. Reidel Publishing Company,
P.O. Box 17, 3300 AA Dordrecht, Holland.

Sold and distributed in the U.S.A. and Canada
by Kluwer Academic Publishers,
190 Old Derby Street, Hingham, MA 02043, U.S.A.

In all other countries, sold and distributed
by Kluwer Academic Publishers Group,
P.O. Box 322, 3300 AH Dordrecht, Holland.

Also published in 1983 in hardbound edition by Reidel in the series Episteme, Volume 10

Printed in The Netherlands

TABLE OF CONTENTS

SECTION VI. THE LEGACY OF MARX

SECTION VII. THE LEGACY OF SPENGLER

SECTION VIII. SOCIOLOGICAL MATERIALISM AND HISTORY OF MATHEMATICS: AN EXPLORATORY CASE STUDY

PREFACE

The problems I address in this book are among the least studied in the sociology of science and knowledge. Part I is a critique of the claim that there are parallels between ancient mysticism and modern physics, and a sociological analysis of this claim as a strategy in intellectual conflict. This study must ultimately be rooted more firmly in a type of sociology of knowledge that is just now beginning to crystallize (and which I discuss in Chapter 7), and a sociology of religion that is not so much unknown as underground, and timid, that is, a non-worshipful materialist sociology of religion. My study of physics-mysticism parallelism is a vehicle for exploring epistemic strategies. I thus conclude Part I by sketching a materialist, emancipatory epistemic strategy. My conclusion brings together a number of ideas formulated by myself and others over the past several years, but stops short of a systematic synthesis. A more integrated and coherent "model" than what I can sketch here must wait on the results of research now in progress in the critical (as opposed to apologetic or worshipful) sociology of knowledge.

Additional ingredients of a materialist epistemic strategy are presented in Part II, which is a general introduction to materialist sociology of mathematics. It is not intended to be a balanced portrait or history of the sociology of mathematics. My aim is to sketch a particular perspective on mathematics as a social phenomenon. This goal has priority over "self-criticism", but this does not preclude a certain amount of criticism aimed at eliminating purist arguments from materialist sociology of mathematics. The study of Greek and European mathematics in Chapters 14 and 15 is a quite preliminary and very general account, but one that I think clearly establishes a rationale for in-depth sociological studies of the calculus. In the interest of focusing on a certain way of telling stories about mathematics and of staying within the boundaries of a relatively elementary presentation, I have generally avoided raising or discussing problems in critical historiography of mathematics. Against the background of this global treatment, I am currently engaged in a detailed analysis of the social origins and implications of Newton's method of fluxions and Leibniz's monadic calculus, and related studies. Other detailed studies of related problems are being carried out by other sociologists and social historians of mathematics.

I advocate a broadly conceived and flexible materialist sociology of knowledge in opposition to all forms of transcendental, idealist and idealistic, purist, individualist-heroic, and religious, theological, and spiritual analyses and explanations of social life. I cannot claim that I have purged my approach of all the ingredients of non-materialist modes of thought, only that I have sought to do so. And I cannot and do not wish to claim that I have adhered to any purist or authoritative notions of what constitutes a materialist and/or sociological approach. I have used the term "social fact" as a reminder of Durkheim's sociological imperative: social facts are to be explained by social facts. I am opposed, however, to any sociologistic extensions of the Durkheimian imperative; all explanatory modes are limited and problematic, and their power (or utility) is a function of how readily they can be linked to other modes, modified, expanded, or otherwise changed (and perhaps abandoned) in response to new problems and new explanatory goals.

The title of this book reflects the subject matter of two separate studies. But the suggestion that this is a study of the interrelationships among physics, mysticism, and mathematics is in a sense a promise. There are a few hints of that promise in this book; in the most general terms, it is that materialist sociology can reveal the common social and motivational roots of science and religion. Materialist sociology can sever in reality ties between science and religion that so far have been cut for the most part only rhetorically, or half-heartedly. This, of course, is only one part of a broader problem: disconnecting inquiry from politics and economics as-they-are, and connecting it to liberatory social relationships.

It is inevitable that in a multi-disciplinary effort of this kind and this scope there will be more than the usual number of vulnerable assertions and errors for readers to gather up. This feature is probably exaggerated because this book is not only made up of two parts, but also encompasses a number of lines of inquiry that I have pursued somewhat independently over the past several years. In the end, I hope any shortcomings will not obscure the power and promise of materialist sociology as an aid to understanding "esoteric" knowledge, and as an epistemic strategy. My conjectures are invitations to criticisms and further inquiries into the social relations of physics, mysticism, and mathematics.

ACKNOWLEDGEMENTS

Part I of this book is a slightly expanded and edited version of a two-part article originally published in *Social Studies of Science* (a Sage publication): 'Parallels and Paradoxes in Modern Physics and Eastern Mysticism, I – A Critical Reconnaissance', appeared in Vol. **8**, No. 2, May 1978, pp. 143–181; 'Parallels and Paradoxes in Modern Physics and Eastern Mysticism, II – A Sociological Perspective', appeared in Vol. **12**, No. 1, February 1982, pp. 37–71. Acknowledgements for other previously published materials used in Part I are provided in the Notes. In Part II, I have drawn on the following previously published materials: Raymond L. Wilder, *Mathematics as a Cultural System,* Pergamon Press, 1981; 'Marxism and Mathematics: The Struggle with Mannheim's Ghost', a review essay on A. Sohn-Rethel, Intellectual and Manual Labor, in S. Restivo (ed.), special issue on 'New Directions in the Sociology of Science', *Newsletter of the International Society for the Sociology of Knowledge*, Vol. 7, Nos. 1 & 2, May 1981, pp. 33–35; 'Mathematics and the Limits of the Sociology of Knowledge', *Social Science Information*, Vol. **20**, Nos. 4 & 5, 1981, pp. 679–701; 'Mathematics and the Sociology of Knowledge', *Knowledge: Creation, Diffusion, Utilization*, Vol. **4**, No. 1, September 1982, pp. 127–144; and 'Mathematics and Civilization' (co-authored with R. Collins), *The Centennial Review*, Vol. **XXVI**, No. 3, 1982, pp. 277–201; 'laws' from Michael J. Crowe, "Ten 'Laws' Concerning Patterns of Change in the History of Mathematics", *Historia Mathematica* 2, 1975, 161–166; and 'laws' from Raymond L. Wilder, *Evolution of Mathematical Concepts*, Halsted Press Book, 1975, pp. 199–201, and pp. 56, 99, 130, 171.

I am grateful to the publishers of previously published materials for permission to publish those materials in this book, and to Daryl Chubin, Randall Collins, and Michael Zenzen for permission to include a few selections from co-authored works. I am indebted to Daryl, Randy, and Mike for far more than is evident in the papers I have written with them. Among the many others who have provided criticisms, suggestions, and other sorts of support related to the preparation of this book, I want to especially thank Valerie Alia, and Irving Adler. I also want to thank Mario Bunge for his consistent support during the various stages leading to the publication of this book in the Episteme series. Finally, I want to thank Sandy Dahoda for her invaluable assistance in preparing the manuscript for publication.

PART I

PHYSICS, MYSTICISM, AND SOCIETY

SECTION I

THE PARALLELIST PERSPECTIVE

CHAPTER 1

PARALLELS BETWEEN PHYSICS AND MYSTICISM

INTRODUCTION

Several years ago Jacob Needleman reported that a physicist was attempting to "correlate and complement" modern physics with the Vedas and William Blake's poetry, another physicist was drawing parallels between quantum mechanics and the teachings of the Tibetan lama Chogyam Trungpa, and a third physicist was correlating Hinduism and Transcendental Meditation with the principles of physics. At that time, the thesis that there are parallels between physics and mysticism ("parallelism") was becoming a strong counterpoint to the counterculture critique of science, and to anti-science and anti-religion sentiments and movements.[1] Parallelism is an important intellectual current at the interface of science and religion, theology, and mysticism. The nature of parallelism is, however, unclear. Parallelists have not reflected critically on the nature and origins of their claims. At the same time, students of science and society have not been attracted to the study of parallelism. One reason for this neglect, perhaps, is that until recently parallelism has been (a) peripheral to the mainstream history of ideas, nourished to a great extent in the literature of the occult- and pseudo-sciences, and (b) characterized by extravagant claims such as the thesis that the central ideas of modern science are presaged in the *I Ching*, and the argument that Einstein's theories were anticipated by one of the Yellow Emperor's advisors forty-five hundred years ago.[2] Such claims are often based on misinterpretations and mistranslations of misdated texts.[3] This situation has changed over the past several years, and parallelist arguments are increasingly being presented and detailed in scholarly as well as popular forums. The synthesis of mysticism and science is also a guiding idea in educational-spiritual communities such as the Lindisfarne Association.

Lindisfarne is named after the medieval monastic school founded by St. Aidan. The founding director of Lindisfarne is W. I. Thompson. According to Thompson, the "parents of the modern world were the Protestant Ethic and the spirit of industrial capitalism". Benjamin Franklin was one child of the marriage between Protestantism and Capitalism. Today, a "New Age" is emerging. The "New Age" Franklin is Buckminister Fuller. Thompson argues that science and mysticism are the parents of the "New Age". The Lindisfarne Association, along with Fuller, astronaut Russell Schweikart, architect Paolo Soleri, and composer Karlheinz Stockhausen are expressions, according to Thompson, of the "marriage of mystical religion and Pythagorean science".[4]

The credibility of parallelism has been strengthened by the observations of eminent scientists on complementarity and convergence between science and religion in general, and between physics and mysticism in particular. J. Robert Oppenheimer, Niels Bohr, Werner Heisenberg, and C. H. Townes are among the many scientists who have lent support, directly and indirectly, to parallelist ideas.[5]

The growing interest, advocacy, and activity focused on parallelism raises numerous sociological problems. Such problems cannot, however, be addressed without first considering problems which arise because of the generally uncritical nature of the arguments for parallelism. Section I of this book is, therefore, a critical introduction to contemporary physics-mysticism parallelism, and the pitfalls of parallelist claims. It constitutes a prolegomenon to the sociological study of parallelism in two respects. First, it is an introduction to the literature on parallelism. Second, it draws attention to some of the pitfalls of parallelism which must be taken into account by sociologists interested in the relationships between physics and mysticism, and in the social origins and functions of parallelism.

I focus on physics-mysticism parallelism. One reason for this focus is that physics-mysticism parallelism juxtaposes the paradigmatic science and the paradigmatic spiritual experience, commonly supposed to be opposing modes of consciousness and knowing. Within physics, elementary particle theory has received the special attention of parallelists. This reflects the manner in which this field is characterized by some distinguished physicists. Jean-Pierre Vigier, for example, describes elementary particle theory as the most modern theory in high-energy physics, among the most advanced of all physical theories, and the theory most relevant to the experimental validation of our assumptions about the deep nature of matter. This view, coupled with speculations by some physicists on the relevance of consciousness to descriptions and interpretations of atomic phenomena, could help to explain why physicists have been drawn to explorations of states of consciousness – including mysticism – and why, simultaneously, mystics and students of consciousness have been listening to physicists. There are, of course, earlier examples of this sort of communication and collaboration, notably the relationship between the physicist Wolfgang Pauli and the psychologist Carl Jung.[6] There are other related reasons which justify an emphasis on physics-mysticism parallelism, and the assumption that this will bring the problems and prospects of the parallelist thesis into the sharpest focus: (a) even in generalized versions of parallelism, science is often a synonym for physics, especially quantum and/or relativity theory; (b) recent developments in parallelism have centered around the physics-mysticism version; and (c) one of the most interesting developments in parallelism is the attempt to draw a connection between the "bootstrap" hypothesis in high-energy physics and Eastern mysticism.

PARALLELISM IN PERSPECTIVE

Parallelism is not a new phenomenon. Before examining the specific version which is the subject of this book, I want to briefly note the place of parallelism in studies of science and religion. Later, in Section III, I will discuss the place of parallelism in society and history.

Ian Barbour has distinguished three basic ways of viewing the relations between science and religion. One view is that science and religion are radically different. The three basic orientations in this "contrasts" perspective are (a) non-orthodox theology, which emphasizes the uniqueness of revelation; (b) existentialism, which distinguishes science and religion in terms of the distinction between personal existence and impersonal objects; and (c) linguistic analysis, which emphasizes the functional differences between scientific and religious languages. A second view is that the existence of God can be inferred from scientific discoveries. This view has its historical roots in the harmony between science and religion expressed in the works of the seventeenth-century "virtuosi", including Robert Boyle and Isaac Newton. Modern representatives of this "natural theology" include Arthur Compton and Arthur Eddington. The third view identified by Barbour and championed by liberal theologians and process philosophers emphasizes general methodological parallels between science and religion. Liberal theologians, for example, claim that science and religion in general are characterized by empiricism, rationalism and the critical interpretation of human experience. They also claim that science and religion are both characterized by presuppositions and moral commitments. It is interesting to note that the elements of A. N. Whitehead's process philosophy appear in the later arguments for physics-mysticism parallelism, particularly in the bootstrap-Buddhism thesis which I will discuss in some detail. These elements include a view of the world as a process in becoming, reality as a set of interconnected events, a relational conception of things, organicism (as opposed to mechanicism), and the self-creation of events.[7]

There are other versions of parallelism which vary in strength and generality. There is a relatively weak version in which science and Taoism are considered "supplementary". Stronger versions emphasize complementarity, or convergence. Parallelist claims can also be classified by the extent to which they refer to science and/or religion. Some authors compare science and religion; some relate Taoism to science. Bootstrap physics has been correlated with Mahayana Buddhism. And one author argues that the world views of physicists and mystics cannot be distinguished. The following outline of physics-mysticism parallelism draws primarily on F. Capra's *The Tao of Physics*, an essay on modern physics and Eastern mysticism.[8]

PHYSICS AND MYSTICISM

When the physicist Niels Bohr, author of the complementarity principle, was knighted in 1947 he choose for his coat-of-arms the familiar yin-yang symbol (t'ai-chi), and the inscription, "Contraria Sunt Complementa" – opposites are complementary. Capra seizes on these choices as (a) an acknowledgement by Bohr of the harmony between Eastern wisdom and Western science, and (b) a dramatic symbolization of his thesis that the principal ideas of modern physics "confirm", "rediscover", or otherwise parallel the ideas of Hinduism, Buddhism, and Taoism. Whether, in fact, Bohr's acknowledgement reveals a "true" parallel between complementarity and yin-yang is one of the issues addressed in this book.[9]

Capra argues that there are parallels between modern physics and mysticism. He focuses on Eastern mysticism because of its central place in Eastern culture; Western mysticism, by contrast, is a marginal feature of Western culture. Capra's referent for "mysticism" broadly encompasses religion, philosophy, thought, and wisdom; in fact, he uses these terms synonymously. It will be worthwhile to digress briefly and to consider some other perspectives on Eastern and Western mysticism which suggest complexities Capra ignores or dismisses.[10]

Conze argues that up to about AD 1450 a "perennial philosophy" dominated East and West. Its basic doctrines were (a) there is a hierarchy of persons, some of whom can know more than others; (b) there are levels of reality, some more "real" than others because they are more "exalted"; (c) "true reality" is accessible to persons with extraordinary faculties through, for example, "prajna" (Buddhism), "logos" (Parmenides), "sophia" (Aristotle), "amor dei intellectualis" (Spinoza), and "Vernunft" (Hegel); and (d) "true teaching" rests on the exemplary lives of charismatic exponents. The special qualities of the perennial philosophy in India included yoga, as the root of "worth-while experience", "karma", and reincarnation. After 1450, the West experienced an industrial, scientific-technological revolution which gave rise to a "sciential" philosophy.[11]

Otto distinguishes Eastern and Western mysticism as "the two principal classic types", and points out their uniformities while challenging the assumption that mysticism is "one and ever the same". And Smart contends that Western religion has been predominantly "numinous" while Eastern religion has been predominantly "mystical".[12] I will return to some of these issues in Section II when I sketch a sociological account of mysticism.

CAPRA'S PARALLELISM THESIS

The basic parallels between modern physics and Eastern mysticism that Capra identifies are (a) organicism, and an "ecological", or "wholistic" view of reality; (b) paradoxes (for example wave-particle duality in physics, koans in Zen Buddhism) that resist logical resolution, appear to be inherent in non-ordinary reality, and are fundamental sources of insights; (c) transcendence of ordinary language and reasoning, and of traditional ideas of space, time, isolated objects and events, and causality (the worlds of the very small, the very big, and the very fast in physics, and the states of higher consciousness in mysticism are on a non-ordinary level of reality); (d) space-time (mystics, according to Capra, have an intuitive comprehension of the modern relativistic concept of four-dimensional space-time); (e) oneness, the idea that the infinite variety of things in the universe manifests one ultimate reality (referred to as Brahman by Hindus, Dharmakaya by Buddhists, Tao by Taoists, and the unified field equation by physicists); and (f) empiricism, manifested as reliance on experimental methods in physics, and on meditative insights in mysticism (variously referred to as "watching", "looking", and "seeing" rather than "thinking"). In addition, Capra identifies the following "equivalencies" in, respectively, modern physics and Eastern mysticism: (a) the quantum field and ch'i; (b) the "physical vacuum" and the Great Void; (c) *S*-matrix theory and the I Ching; (d) complementarity and Tao; and (e) the bootstrap model in high-energy physics and Buddhism: this equivalency will be discussed below.

The evidence for these parallels consists of statements from primary and secondary sources on how Hindus, Buddhists, Taoists, and physicists "see" reality. Capra admits that such statements cannot serve as rigorous demonstrations for his thesis. He relies on them, however, to stimulate an appreciation of parallelism as a subjective experience. This is Capra's objective in juxtaposing statements on physics and mysticism and drawing attention to common images of reality. Two illustrative examples are: (a) the metaphor of the cosmic dance, and (b) statements on the physical vacuum, and ch'i.

The familiar bubble chamber photographs from the laboratories of high-energy physicists are manifestations of the complex world of elementary particles. That world is, theoretically, more complicated than the patterns observed in the bubble chamber because certain particles which are exchanged in particle interactions – so-called "virtual particles" – are not observed. Thus, the Feynman (network) diagrams that schematically represent particle interactions are more complex than the patterns observed in the laboratory because they take virtual particles into account. Capra quotes physicist Kenneth Ford's comments on such a diagram: "Every proton occasionally goes through exactly this dance of creation and destruction".[13] Such dance imagery, Capra points out, is not uncommon in modern physics; "energy dance", for example, is another term used to describe particle interactions. Dance imagery is also important in mysticism. Capra refers in particular to one of the most impressive examples of dance imagery from the perspective of parallelism, The Dance of Shiva. Shiva is the multi-armed Hindu god whose dance of creation and destruction is captured in a wide variety of bronze sculptures dating from the tenth century. The Dance of Shiva conveys the Hindu conception of reality as a continuous process of creation and destruction. According to Capra, this parallels the modern physicist's conception of self-interacting particles emitting and absorbing virtual particles in an energy dance that is the foundation of physical reality.

Historically, dance imagery is not uncommon in mythical, religious, philosophical, and scientific discourses on reality. The dance imagery used by some modern physicists has precedents. Leonardo da Vinci, for example, depicted the cosmos as an astral ballet, and wrote that "Movement is the cause of all life".[14]

The parallelist method of juxtaposing quotations is illustrated by the following statements on quantum field theory and ch'i; the first is by the physicist W. Thirring, the second by the eleventh century Chinese sage, Chang Tsai:[15]

(a) The field exists always and everywhere; it can never be removed. It is the carrier of all material phenomena. It is the "void" out of which the proton creates the pi-mesons. Being and fading of particles are merely forms of motion of the field. (b) When one knows the Great Void is full of Ch'i, one realizes that there is no such thing as nothingness.

Ch'i (which Needham describes in modern terms as "matter-energy") is one of the two fundamental concepts underlying the neo-Confucian view of the universe; the other is *li*, "the principle of organization and pattern in all its forms". Ch'i is one of the fundamental ideas in the Taoist literature too, and has been variously translated as "passion nature", "material principle", "constitutive ether", "force, energy, breath, power", "ether or force", "the great breath of the universe", and so on. R. G. H. Siu notes, in his own words, Chu Hsi's (1130–1200) elaboration of Chang Tsai's notion of the condensation of ch'i: "The universe is formed by the ch'i, as activating essence, putting the li, as ultimate reason or purpose into concrete form." In his own exercise in parallelism, Siu contends that "Light itself consists of energy and ch'i. Quantum properties of light are the refractions of its mass-energy component. Continua are the refractions of its massless ch'i. It's no wonder that a Sanskrit root for Time is Light".[16]

The following description of Capra's case study of bootstrap and Buddhism provides a detailed account of the methods and theses of parallelism.

BOOTSTRAP AND BUDDHISM

The bootstrap hypothesis, formulated by Geoffrey Chew, is that there are no fundamental entities (for example laws, particles, fields, principles, equations) in nature.[17] Capra considers this hypothesis to be the culmination of changes in our view of nature initiated by the introduction of the interrelational concepts of quantum theory, developed further in the relativistic explanation of quantum dynamics, and carried further still by the introduction of reaction probabilities in *S*-matrix theory.

S-matrix theory is one of the three theories which has guided work in elementary particle physics. The other two are field theory and group theory. J. A. Wheeler defined the *S*-matrix in 1937, and in 1943 Heisenberg suggested it might be able to play a fundamental role in quantum physics. Their inability to compute interparticle forces led Heisenberg and other *S*-matrix students to abandon this line of inquiry. Gell–Mann, Landau, and Mandelstam later revived interest in the theory. Chew was one of the persons responsible for freeing *S*-matrix theory fully from its link with field theory by adding a postulate of analyticity to Heisenberg's *S*-matrix notions.[18] The "*S*" in *S*-matrix stands for "Streu", the German word for "scatter". The mathematics of the *S*-matrix is elementary (the standard theory of analytic functions of complex variables) in contrast to field theory (the most mathematically rigorous and physically remote of the three theories) and group theory.

S-matrix theory relies heavily on experimental data; it is so closely tied to experiment, in fact, that some critics have dismissed it as contentless "sophisticated phenomenology".[19] The goal of the *S*-matrix theory of strong interactions is, " . . . given certain symmetries, to predict all the observed particles, together with masses and mutual interactions, in terms of a single constant with the dimensions of length". No arbitrary dimensionless constant should be introduced. Technically, this goal is stated in the form of a postulate: "The *S*-matrix is a Lorentz-invariant analytic function of all momentum variables with only those singularities required by unitarity".[20] These formal ideas are reflected in the formulation of the bootstrap model that follows.

The basic characteristics of the bootstrap model are: (a) a conception of nature as a self-consistent whole (that is, all of physics is conceived to follow from one requirement – its components must be consistent with themselves and one another, thus eliminating, ideally, the need for arbitrary, in the sense of unexplained, parameters such as the fundamental constants); (b) an approximation that neglects electromagnetic, weak, and gravitational interactions and focuses on the strong interactions (the rationale for this approximation being related to the fact that the strong interactions are stronger by many orders of magnitude than the others: since the strongly interacting particles are known as hadrons, this approximation is called the "hadron bootstrap"); (c) hadrons are composites; hadrons are constitutents of hadrons; and hadrons are a binding force between hadrons (based on the conception of a force between two particles as an exchange of other particles; in this case the exchanged particles are conceived to be hadrons); the name "bootstrap" is derived from this picture of a set of hadrons which generates itself, that is, "pulls itself up by the bootstrap"; (d) the hadron dynamism is "circular and violently nonlinear", with no free parameters; it is self-determining (that is, it can be realized in one, and only one way), and therefore the only possible set of hadrons is the set in nature; (e) hadronic properties follow uniquely from self-consistency (bootstrap), and do not have to be introduced as fundamental quantities (that is, as "basic building blocks of nature"); following Gell–Mann's suggestion, this idea is referred to as "nuclear democracy", and is opposed to "aristocratic atomic physics". "Nuclear democracy" describes the notion that hadrons are composite, or bound, states which exist entirely by virtue of Yukawa-type forces. In *S*-matrix terminology, each nuclear particle in a dynamic-governed nuclear democracy " . . . is conjectured to be a bound state of those *S*-matrix channels with which it communicates, arising from forces associated with the exchange of particles that communicate with 'crossed' channels . . . ".[21]

In contrast to bootstrappers, fundamentalist high-energy physicists are committed to the search for basic building blocks. They accept the traditional view of fundamental constants (for example, the velocity of light, the mass of the electron, and the magnitude of the electron's charge) as empirically determined and theoretically unexplained. The value for the velocity of light, for example, is not explained in relativity theory. Bootstrappers argue that, ideally, all so-called fundamental constants should be explained in terms of the self-consistency hypothesis. Feynman, expressing the fundamentalist view, describes the bootstrap notion as something " . . . that sounds like wagging the dog by the tail".[22]

Capra argues that the fundamentalist view is rooted in the ways of Western science, which in turn reflects traditions that spring from the Greek heritage; the bootstrap model, by contrast, is more consistent with Eastern modes of thought. In particular, Capra stresses the parallels between the bootstrap model and Mahayana Buddhism, which is later, more complex, and more other-worldly than Hinayana Buddhism. Capra follows Buddhist scholars such as D. T. Suzuki and E. Conze in viewing Mahayana Buddhism, represented by the Avatamsaka Sutra, as the culmination of Buddhist thought, sentiment, and experience. More specifically, Capra, following Suzuki and Chang, considers Mahayana Buddhism (especially in the Hua-Yen school of China) to be the fullest elaboration and clearest expression of the Buddhist idea that all things and events are interrelated and part of a "unity" or "totality".[23]

The two principal schools of Buddhist thought are the Hinayana and the Mahayana. Hinayana Buddhism, the "orthodox" branch, became established in Ceylon (now Sri Lanka), Burma, and Thailand. Mahayana Buddhism developed in Nepal, Tibet, China, and Japan, and is generally considered the more important and philosophically interesting of the two schools. The culmination of Buddhist thought in the Mahayana school occurs, according to Suzuki, in the Hua-Yen school created by Chinese translators and interpreters of the Avatamsaka Sutra during the T'ang Dynasty. Hua–Yen (or Hwa-Yen) means "flower decorations", or "garland"; the Avatamsaka Sutra is known as the Garland Sutra. An important stumbling block for parallelists is the problem of deciding which of the numerous presentations of the Buddha's thought is the "most authentic" – the Buddhism of the Pala period, Yogacarins, the Pali Canon, the scholastic Hinayana and the work of the later logicians, early Mahayana and Zen, the Saddharmapundarika, and so on.[24]

From Capra's parallelist perspective, the most important idea in the Avatamsaka Sutra, especially with respect to the bootstrap model, is "interpenetration". This idea is expressed in the 2,500 year old metaphor of "Indra's net". Indra is a Hindu god who appears as a protector of Buddhism in many sutras. His "heaven" (or the cover over his "celestial palace") is portrayed as a network of jewels arranged in such a way that looking at any one you can see all the others reflected in it. This metaphor reminds us that "In every particle of dust, there are present Buddhas without number". This, according to Capra, is the same non-ordinary idea bootstrappers have of hadronic reality. Following a review of the various formulations of the bootstrap idea including the technical *S*-matrix formulation, Capra concludes that bootstrap can be " ... summed up in the provocative phrase, 'every particle consists of all other particles' ". In Chang's description of the "net" metaphor, mutual reflection forms a "realms-embracing-realms-net" within each jewel or pearl. According to Chang, the metaphor illustrates " ... the mutual penetration and the containment of Totality of the Hwa Yen Dharmadhatu". "Dharmadhatu" refers to the Infinity and Totality of the Domain of Buddha.[25]

In concluding his argument for parallelism, Capra speculates on the possibility of going beyond the transformation of the scientist from observer to participant wrought by modern physics. Chew himself, in fact, conjectures that by logical extension the bootstrap model implies that the existence of consciousness is necessary for self-consistency in the universe. This points the way toward explicitly incorporating consciousness into physical theories, a speculation endorsed by Wigner. This entails the transformation of the bootstrap model into a vision transcending ordinary realms of thought and language, bringing science into the Mahayana world of the unthinkable, "acintya".[26]

Having outlined the basic ideas of physics-mysticism parallelism, I want to turn now to a consideration of pitfalls in the data, methods, and assumptions of parallelism, and to a fuller exploration of the comparisons Capra and other parallelists rely on in their arguments.

The Appendix to this chapter is a report of an encounter between Capra the physicist and Pir Vilayat the Sufi master. Their meeting realized in life the meeting of alternate realities Capra discusses in his book.

APPENDIX A

THE PHYSICIST AND THE SUFI MASTER:

A Meeting of Alternate Realities

Fritjof Capra published *The Tao of Physics* in 1975. Since then, he has been lecturing widely in the United States and other countries on the thesis he defends in his book: there are parallels between modern physics and Eastern mysticism, and these parallels suggest a new wholistic world view opposed to the mechanistic, fragmented world view that prevails in modern society. "The survival of our whole civilization", Capra argues in the closing paragraph of his book, "may depend on whether we can bring about such a change. It will depend, ultimately, on our ability to adopt some of the *yin* attitudes of Eastern mysticism; to experience the wholeness of nature and the art of living with it in harmony".

Capra carries buttons showing the familiar "Diagram of the Supreme Ultimate", widely known as the symbolic representation of yang-and-yin. The light yang side is filled with physics equations; the dark yin side shows part of the Buddha's face, from a fifth century AD Indian sculpture. Capra has taught at the University of California in Berkeley, and done research in high-energy physics at the Lawrence Berkeley Laboratory.

On June 10 and 11, 1978, Capra had his first opportunity to enter into, in his words, "a full exchange with a representative of the mystical tradition". The representative was Pir Vilayat Inayat Khan, Head of the Sufi Order in the West, and eldest son of Hazrat Pir-O-Murshid Inayat Khan, founder of the Sufi Order. The setting was The Abode of the Message in New Lebanon, New York, one of nearly one hundred Sufi communities in the United States, Puerto Rico, and Canada, and national headquarters of the Order. In another time, this was the administrative center of the American Shaker community. The members of the Abode live communally, many in Shaker-built buildings, in the shadow of Mount Lebanon. They have developed ties to the local community through a variety of enterprises, including Heart & Wings, a VW repair shop in New Lebanon, and the Mountain Road Children's School in Canaan, New York. The Abode has developed a program of seminars, workshops, courses, and retreats under the aegis of the Omega Institute. Capra and Pir Vilayat met in the first Omega seminar, held in the Meditation Hall overlooking the Abode farm.

I was among about eighty people who gathered in Meditation Hall on the first morning to hear Pir Vilayat. Pir described himself as a "contemplative" interested in what science has to say about the contemplative's experiences. Occasionally, he would draw a sheet from the small pile of notes at his side to read a quotation on wholism, holograms, synchronicity, or the transcendence of space and time. The scientists he quoted hold views which can broadly be described as "wholistic": the mathematician Charles Musés, the neurophysiologist Karl Pribram, the quantum physicist David Bohm, and – from an earlier generation – Wolfgang Pauli, and Werner Heisenberg. Leibniz, a seminal figure in the pantheon of wholistic science, was called on too, as was the psychologist Carl Jung.

Mystics, Pir said, are continually reaching beyond personal consciousness. But, he added (and here he parts company with many mystics), the individual should never be totally absorbed in the All: this is Pir's principle of "Samadhi With Open Eyes".

One dimension of our self is localized; another is "spread throughout the universe". This is akin, Pir argued, to the particle-wave duality in physics. There is a personal vantage point (the "particle view of things") and a cosmic vantage point (the "wave view of things"). The Sufi strives to experience both the personal and the cosmic, the world of the lens (which helps us to objectivize reality, to separate "subject" and "object") and the world that transcends the lens. If the lens is relevant to the reality of everyday life, the hologram is relevant to what lies "behind" or "beyond" that reality. Pir described the hologram as a "democratic structure" opposed to but interdependent with a "hierarchical structure". The hologram is important for Pir (as it is for Pribram and Bohm) because of the extent to which the *whole* holographically photographed object can be reconstructed from a *part* of the holographic plate. Pir calls the experience of reality in terms of *both* the lens and the hologram "stereoscopic consciousness". To achieve this is to "awaken". This is the Sufi Message: "a call to those whose hour has come to awake, and . . . a lullaby to those who are still meant to sleep". The Message is communicated in words and in silence. Pir reminisced about his earlier days "on the guru hunt": "I always enjoyed", he said, "being in the presence of people who did not speak".[27]

The root of the spiritual experience, as described by Pir, is an "awakening" in which the physical world is "out of focus". The contemplative's awakening from "*a* perspective" – from "*everyone's* perspective" – is paralleled by the break with causality in modern physics; in Pir's words, "middle-range" or "supermarket" reality is transcended. Jung's idea of synchronicity, Pir noted, captures the essence of what it means to transcend causation and to become aware of (awake to) the interrelationships among events.

The mystic is "always experiencing the *essence* behind all things", according to Pir. He referred to this as "the deeper sense", and explained it in terms of the difference between the "sense" of the tongue and the "deeper sense" of the stomach, which will "reject" certain things the tongue "accepts". The deeper sense opens us to "the symphony of the universe", a "pure experience of the universe without matter", the "eternal and the unchanging".

Pir noted that insofar as the essence of spiritual life is an ever-present consciousness of the whole, mystics experience the same reality that schizophrenics do. Unlike schizophrenics, however, mystics are able to integrate ordinary and extraordinary experiences. Later in the seminar, Capra described a conversation with R. D. Laing in which the psychiatrist said that "mystics and schizophrenics are in the same ocean, but the mystics swim while the schizophrenics drown".

Scientists, Pir believes, are moving toward an interpretation of reality that takes account of causal and acausal (synchronous) experiences and explanations. He modestly but unhesitatingly "stumbled about in physics", and was most dependent on his notes when describing scientific concepts. He openly welcomed the opportunity to have a physicist sitting next to him who could correct him. This was the cue for Capra, who had quietly entered the Hall shortly before the end of Pir's lecture, and settled himself into a half-lotus position next to Pir.

Capra began his lecture by outlining some of the revolutionary changes occurring in our society. Two modes of consciousness have been identified in various cultures and given a variety of names: scientific and mystical, rational and intuitive, yang and yin. Modern industrial society has favored yang consciousness and is "out of balance". An evolutionary movement toward balance is reflected in the strong interest evident today in mysticism, psychic phenomena, ecology, wholistic healing, and feminism. All of these things are interrelated aspects of a "cultural revolution". Feminism, according to Capra, may be the most crucial feature of the revolution; women are playing a central role in bringing about changes that spell the end of patriarchal culture. This is the context for the dialogue between science and mysticism.

Capra pointed out that science and mysticism have been separated in the development of human thought, and now represent extreme specializations, respectively, of the rational and intuitive modes of consciousness. In fact, however, they complement and supplement each other. Science (particularly physics), Capra argued, is moving towards wholism and revealing truths known to the mystics of all times and ages. Most scientists are unaware of this trend; in fact, said Capra, "it goes against their will".

The cultural revolution described by Capra is carrying us from a fragmentary to a wholistic world view. This process is linked to an increasing awareness of ecology, the interdependence of all things, and the need for global unity. Mystical truths are being recognized as ideas are exchanged across cultural boundaries, and as economic and social structures are transformed.

Capra expressed disappointment with many representatives of Eastern religions. "They are often part of the problem", he said, given "their real estate holdings", and "the way they treat women". Their actions contradict the truths uncovered in ancient Hinduism, Buddhism, and Taoism and upheld by all of the true mystics.

Science is grounded in everyday experience, intellect, abstractions, linear thinking, and language. Following the philosopher Korzybski's map-territory metaphor, Capra said that we can become trapped into believing that our words and theories (the maps) *are* reality (the territory). One of the main functions of Eastern wisdom is to free us from this trap.

Scientific understanding, Capra argued, is ultimately rooted in words and experiments – in the verbal interpretation of mathematical models, and in the experimental validation of theory. Mystical (or intuitive) understanding is based on experience. Pir pointed out, however, that there is something in mysticism that cannot be described as "experience"; this is "the light thrust upon all things", "knowledge within itself", "knowledge *in you*". It is knowledge that comes without "data processing". Capra replied that perhaps the reality within and the reality without are ultimately identical.

One objection to Capra's conception of parallels between science and mysticism is that science is tied to external, technological "things", whereas mysticism involves "looking within". A second objection is that science operates in the public domain, whereas mysticism is a private, even a *privatized*, experience. But Capra argued that these are superficial differences. He pointed out that "the real scientists" are motivated by those rare moments of intuitive flashes (insight) which are an acknowledged part of the discovery process in science. It is precisely at this point, Pir interjected, that mysticism enters the picture. This is neglected in teaching science in our fragmented society.

Scientists are not interested in Truth, Capra asserted, but rather in developing reasonably good descriptions of the world. Newtonian mechanics was not wrong; rather, it was limited. Capra asked Pir about the role of *approximations* in mysticism. Pir replied that there are levels of reality, and the contemplative's experience is never absolute. The way the world appears to you depends on your level of consciousness, and is therefore always relative:

Capra: "Don't mystics say there is a level where you are not limited"?
Pir: "Yes, limitlessness. You *are* the totality, not *part* of the totality".
Capra: "The mystic experiences the interconnectedness of all things directly, then".
Pir: "I beg to differ – you can't know everything in mysticism".

Here, as at other times during the seminar, Pir stressed the concepts of "Samadhi With Open Eyes", and relativity: in his view, limitlessness does not mean total absorption of the self, nor does it mean absolute knowledge.

Capra then discussed the *indirect* nature of observation in atomic and subatomic physics. Both mysticism and modern physics transcend ordinary experience and the ordinary language *derived* from that experience. Relativity theory and mysticism, for example, transcend ordinary space and time. Neither lends itself to easy description in ordinary language.

The ecological, wholistic, mystical notions of interconnectedness and interdependence which play a central role in modern physics are opposed to the mechanistic view of classical science in modern physics. But unlike Pir, Capra does not believe that there is "purposiveness" in nature. Pir sees things in terms of the static and dynamic aspects of God, and a "goal-orientation" in the universe. Another and related area of controversy emerged over the idea of "freedom". Pir's argument for "free will" is associated with his feeling that the human self should not be lost in the whole. Capra concluded, in friendly opposition, that in terms of a wholistic, natural science view of reality, the idea of free will must ultimately be abandoned. The concept of "time" was one of the crucial points of controversy in their dialogue on freedom and purposiveness:

Pir: "I cannot conceive how the curvature of space can be equated with time".
Capra: "Space *and* Time are curved."
Pir: "Feynman's notion of particles moving backward in time is just a convenience; things are not really like that".
Capra: "The issue of an 'arrow of time' is not yet resolved".
Pir: "There *is* purposiveness in the universe; but most of the mysticism quotes in your book selectively emphasize the lack of direction in the universe".
Capra: "Yes, I emphasize Taoism, and Buddhism. Time *can* have an arrow in some *section* of the universe".
Pir: "I am not thinking of a Prime Mover, but there *is* something else. I explain Feynman's idea in terms of a different dimension of time,

not the reversal of time. The 'reversal of time' is only a convenient way of speaking".

Capra: "It is *very* convenient"!

Pir: "There are other dimensions which you do not discuss in your book".

Capra: "In science, the linear notion of time is often useful. In particle physics, if you reverse time, it is useful. So maybe we can drop the notion of time and use instead the notion of order".

Pir: "This is exactly what we experience in meditation – evolutionary, transcendental time, eternity. We experience *pure* vibration. We are glad when there is a parallel between science and our experience as contemplatives".

Capra concluded his lecture on the differences between classical mechanistic physics and modern wholistic physics with a set of what he called "very pretty slides" of high energy particle showers that drew "ohs" and "ahs" from the audience. He compared these to the Hindu deity Shiva's dance of creation and destruction.

Capra specializes in *S*-matrix theory. The recent incorporation of the notion of "order" into the theory has helped it to meet the challenge of the *quark*. Quarks – considered by some physicists to be the ultimate constituents of matter – can, according to Capra, be brought into the *S*-matrix using the concept of "ordered interconnectedness". This, he said, represents a major advance for the so-called "bootstrap" physicists (who deny the existence of *fundamental* particles or laws of nature) over the "fundamentalists". More generally, this strengthens the wholistic approach emerging in modern physics.

The dialogue between Pir Vilayat and Capra on evolution was confusing; they both admitted to a limited knowledge of biological science. Nonetheless, they engaged in a very lively speculative dialogue on freedom, genetics, entropy and negentropy, and mutation. The following exchange, which elicited appreciative laughter from the audience, may very well have been the highlight of the session:

Capra: "Your view of DNA in hologram-like information terms is not the majority view in biology today".

Pir: "I know, but it is the position *I* am adopting, since it is in accord with my contemplative experiences".

Pir Vilayat began the second day's morning seminar by skillfully leading the group of experienced and uninitiated meditators through a twenty-minute meditation. In *cosmic consciousness*, the frontiers between the subjective and the objective world dissolve, and there is a loss of personal identity; in *transcendental consciousness*, consciousness is "lifted" from the material level, and the individual grasps "the immaterial world": this is Divine thought, Divine consciousness, the "Samadhi with open eyes" Pir Vilayat defends.

In the final seminar presentation, Capra argued that the widespread problems of fragmentation in self and society can be traced to the tendency among scientists in all fields to follow the lead of Newtonian and Cartesian (classical, mechanistic, dualistic) physics. Scientists today must recognize, he said, that physics has gone beyond this classical framework. He criticized the use of the classical scientific approach in (a) biology, (b) the "medical model" in medicine, (c) behaviorism and psychoanalysis, and (d) orthodox economics. The "new physics" is not merely a unifying force in contemporary thought, according to Capra; it is also a therapeutic force. It implies a change in our view of reality illustrated by the movement towards wholistic medicine. The basic principle guiding this approach is that people can and should be responsible for their own health – as individuals, as families, and as communities. More generally, this suggests that people can and should be responsible for their own lives – in the negative sense that they should not be at the mercy of authoritarian, alienated and alienating practitioners (whether in politics or medicine), and in the positive sense that they should develop and use their individual and collective resources and abilities to deal with the issues, problems, and challenges of their lives.

CHAPTER 2

THE PITFALLS OF PARALLELISM

INITIAL OBSERVATIONS

Physics-mysticism parallelism refers to the two-fold argument, following Capra's formulation, that (a) "a consistent view of the world is beginning to emerge from modern physics which is harmonious with Eastern wisdoms", and (b) "Eastern mysticism provides a consistent and beautiful philosophical framework which can accommodate our most advanced theories of the physical world".[1] The basic data for parallelism are common language (for example, English) statements on the nature and implications of physics and mysticism that vary in technical content. The methodology of parallelism is the comparative analysis of such statements. Similar rhetoric, imagery, and metaphoric content in such statements constitute the evidence for parallelism. The basic assumption in this approach is that if the imagery and the rhetorical and metaphoric content of statements on physics and mysticism are similar, the conceptual content must be similar, and the experience of reality must also be similar among physicists and mystics. Earlier I cited Capra's comparison of statements by Thirring and Chang Tsai. Perhaps the most ambitious application of this method occurs in Needham's studies on Chinese and modern science, characterized by the continuing discovery of Whiteheadian philosophy, dialectical thought, and anticipatory scientific attitudes, concepts, and methods in Chinese texts.[2] This requires searching for, selecting, and translating materials for comparative analysis. At least three difficulties accompany this requirement. One is that representativeness must be achieved first in selecting a particular piece of literature, and then in selecting a particular word, sentence, or paragraph. No rigorous sampling procedures have guided such selections, nor am I aware that any have been proposed with respect to the specific problem of parallelism. In the case of ancient texts, such procedures are not readily applicable due to fragmentation and corruption, as well as to the uncertainties about what texts existed, and still exist undiscovered. Independent judgements of representativeness based on textual (internalist) and contextual (sociocultural) knowledge could help, but have not been systematically cultivated by parallelists.

A second and more serious pitfall arises in actually comparing statements.

Comparable common language statements have to be derived from the mathematical formalism and specialized language of physics, and from the meditative insights of mysticism. In addition, the literature of and on mysticism will in some cases have to be translated from several common languages. In the case of Buddhism, for example, original texts and major translations are written in Sanskrit, Pali, Tibetan, Japanese, Apabrahmsa, and Chinese. This is further complicated by the problem of juxtaposing statements written at different times, sometimes centuries apart. The problems with respect to Buddhism and other Eastern religio-philosophical traditions have been widely noted.[3] As for physics, most contemporary physicists probably share Feynman's despair of communicating the laws of nature to persons ignorant of mathematics.[4] There may be some, though it is unlikely, who share the view expressed, I believe, by Rutherford, that if a theory is good you should be able to explain it to a barmaid! Einstein falls, perhaps, between these extremes in terms of his interest in and capacity for translating the formalism of relativity into English sentences. Under these conditions, it is perhaps reasonable to expect adequacy (if not perfection) in translation, but naive to assume that the comparability of translations can be taken for granted. Some efforts must be made to arrive at a sense of the validity of the assumption that sentences are comparable. This would necessarily involve consideration of the relationship between culture and cognition.[5] Parallelists have not addressed themselves to these questions. Incidentally, some physicists argue that scientific "understanding" (establishing in our own minds the "compelling reasonableness" of phenomena) depends on framing models and visualization. They contend that however useful the reliance on formalisms and algorithms may be (manipulating operators and matrices, for example), it is not a substitute for understanding.[6]

The physicist's claim that the worlds of quantum mechanics and relativity are mathematical worlds which cannot "honestly" (to borrow Feynman's language) be described or visualized in terms of common languages and ordinary experiences should not be taken lightly. Neither should we dismiss the mystic's ultimate silence about non-ordinary experiences. It would be foolhardy to assert dogmatically that physics and mysticism cannot be put into words, now or ever. In any case, attempts *are* made to put non-ordinary mystical experiences into the words of ordinary experiences; and physicists *do* write and talk about their mathematical universe in highly technical physical terms as well as "everyday" vocabularies. These attempts produce parallels across time and culture. The ways in which ordinary words compromise the non-ordinary experiences of physicists and mystics need to be considered more carefully than they have been in advocating and evaluating parallelism. The "imageless" mystical experience may be "only a fleeting moment of amorphous illumination".[7] Once this has passed, the fact that the mystic has a social heritage, including a language, makes its full force felt – and the experience can be written or spoken about.

The third pitfall in comparing statements is contamination. Modern mystics are at least aware of references to the concepts of modern physics that have filtered into the language of our everyday world, especially those regarding space, time, and causality. Some mystics are well-versed in the mathematical formalism of modern physics. At the same time, modern physicists have come into contact – directly and indirectly – with the ideas of mysticism. Gell–Mann, for example, used the term "eightfold way", paralleling "eightfold way" or "eightfold path" in Buddhism, in his application of group theory to the study of elementary particles. The parallel in this case is purely linguistic.[8]

"Levels of reality" hypotheses are relevant to the issue of translation in parallelism. There is a controversy among physicists concerning the extent to which the nature of a given level of physical reality, expressed in a level-appropriate language, can be coordinated with the nature of another level (or other levels) to give a more general theory of nature. The idea of "level-appropriate languages" makes translation problematic. Indeed, some students of science (for example Feyerabend, and to a certain extent Kuhn) have suggested that translation is virtually impossible. It may, however, make sense to assume (following John Graves) that ontic levels are not completely independent, and that, therefore, cognitive levels are not completely independent. Any given level-appropriate language will contain certain general abstractions and idealizations rooted more in cognitive processes than in external realities. Graves suggests that it is generally reasonable to assume that there are commonalities across levels and that these are reflected in language.[9]

The insistence that each level of reality is characterized by an appropriate language with distinctive terms and relations, or laws, does not imply that there are no common terms or relations in linking level-appropriate languages. Most terms may in fact be cross-level terms applicable to a variety of, and in some cases all, levels. In terms of the parallelism argument, we are faced with three questions: (a) can we consider mysticism and physics to be on different levels of reality; (b) does it make sense to consider these levels of reality "bounded" (that is, part of one reality or realm of reality in which Graves' assumptions apply); or (c) do mysticism and physics operate within distinctly bounded realms of reality ("separate realities") for which there are no cross-level terms and relations? The latter case may hold, for example, if there are two realms of reality, one open to symbolic consciousness and expressible in language, and one open to non-symbolic consciousness and not expressible in language. I will discuss this idea in more detail below.

Finally, parallelists and critics of parallelism have to consider the functions of language. The centrality of language – including the language of mathematics – as a medium of communication about reality is well established in science. Language may be more or less abstract, more or less remote from the world of our senses, but it is used to refer to "reality". Language may have a different function in mysticism. Chang, for example, refers to the "fountain-head" of Prajnaparamita teaching (the perfection of Wisdom, the paramount practice and virtue of the practitioner – Bodhisattva – of Mahayana Buddhism) as "direct seeing", an intuitive experience "intrinsically beyond words and symbols". Yet words must be used to communicate the Bodhisattva's experience to other humans. This dilemma is manifested in the practice of making a statement and repudiating it immediately. The result is a literature brimming with paradoxes and contradictions. If these are interpreted in terms of the function of language in scientific discourse, the "Prajna truth" and mysticism in general appear to be a string of absurdities. If, on the other hand, language in mysticism is applied to transcending the limitations and inadequacies of ordinary experience, then the situation loses its absurdity, but becomes incompatible with the assumptions of translatability and comparability in parallelism. In addition, Chang has drawn attention to the difficulties that have arisen when philosophers have insisted on treating Buddhist concepts such as "anatman" (No-Self) as philosophical concepts rather than as meditational, or therapeutic devices. Conze also stresses the subordination of philosophy to a soteriological imperative.[10] These problems are complicated by the insistence among some Eastern intellectuals that their traditional systems of ideas and beliefs be treated as legitimate philosophies.[11] Parallelists have not attended to such problems, and have thus failed to show why functional linguistic differences are irrelevant to their comparative search for conceptual equivalencies.

Linguistic pitfalls arise specifically in the comparative analysis of verbal accounts of experiences in physics and mysticism. Comparing statements in the absence of experience with physics and/or mysticism obviously heightens the dangers of such pitfalls. But experience in one or both of these realms does not eliminate the pitfalls. The problem of translating experiences into words for oneself and for others remains. This is a rich problem-area for sociologists, but one which demands a rare combination of expertise in sociolinguistics, fluency in physics and mysticism, and experience in the physical and mystical realms.[12]

SPACE-TIME PARALLELISM

One of the frequently cited parallels between modern physics and mysticism is the space-time parallel. Capra, for example, argues that modern physics and Eastern mysticism reflect an awareness of the intimate interconnectedness and interpenetration of space and time. The idea found in the Avatamsaka Sutra that there is no space without time and no time without space parallels the view expressed by Minkowski that space by itself and time by itself are "mere shadows". Capra refers here to comments by D. T. Suzuki, and to a passage from the well-known 1908 lecture by Hermann Minkowski.[13] On the surface, such parallels are hardly deniable. It must be remembered, however, that we are dealing with "supposed translational equivalents". The semantics and syntax of "space" and "time" in Plato or the Buddhist sutras on the one hand, and in Einstein, Minkowski, or Wheeler on the other, are radically different.[14] In physics, the conceptual transformation of "space" and "time" has rested heavily on mathematization. There is no reason to suppose that this process of conceptual evolution has ended in physics. But there is no indication that the non-ordinary experience of space and time in mysticism is any different for modern mystics than it was for the ancients. Mysticism does not seem to hold the same inherent promise of a surprising new conception of space and time that physics does.

The mystic's experience of space-time unity may arise from the suspension of ordinary ways of "attending" to the world. C. O. Evans has characterized ordinary symbolic consciousness as "attention deflection" :[15]

> . . . each experience of an event itself causes attention to switch to the experience of another event, and in turn that experience of an event further causes attention to switch to yet another event in an endless concatenation of attention deflections.

The process of attention deflection, according to Evans, is the source of "meaning". Evans hypothesizes that mystical experiences indicate that it is possible to suspend attention deflection, and to attend to the world in a way he describes as "absorptive" attention. Absorptive attention is non-symbolic consciousness, consciousness that cannot be communicated in symbols. The structure of consciousness is dissolved when attention deflection is inhibited or suspended.

Evans' distinction between symbolic (attention deflection) consciousness, and non-symbolic (absorptive attention) consciousness seems to parallel Chang's distinction between ordinary "one-at-a-time" or "shifting realms" consciousness, and non-ordinary "simultaneous-arising" consciousness.[16] In ordinary experience, human beings attend to one realm of reality at a time, and shift from realm to realm as the need for new perspectives arises. Water is a liquid, or H_2O, or the product of electronic movements – but never all of these at the same time. The Mind of Buddhahood, by contrast, experiences all realms interpenetrating harmoniously in the truths of "simultaneous arising" and "simultaneous non-obstruction". Evans' theory suggests that this experience may reflect a human ability to inhibit or suspend attention deflection. This suggests that the mystical experience of space-time is the product of absorptive attention and is not related to the physicist's experience of space and time; the latter is the product of a *conceptual* development that *may* have been in some way affected by absorptive attention but is not entirely explained by that phenomenon. Mystics may indeed experience space and time as a "four-dimensional continuum" in some sense. It is not clear, however, that their "experience" is conceptually equivalent to the continuum "experienced" by physicists and expressed in their mathematical formalism.

It seems improbable that a mystic would independently arrive at the notion that there is more than one type of structure for four-dimensional space-time; on the contrary, there is something surprising, perhaps, but nothing improbable about the recent construction by two mathematicians, Michael Freedman and Simon Donaldson, of an existence proof for just that idea. It would not be surprising to find that if the idea catches on in science, some mystics or their surrogates will come forward to claim that they knew it all along.

COMPLEMENTARITY PARALLELISM

Nowhere are the pitfalls of parallelism better highlighted than in "complementarity parallelism", perhaps the most widely discussed and debated version of the thesis. Complementarity parallelism is based on the idea of complementarity in quantum theory, introduced by Bohr at a 1927 conference in Como, Italy, commemorating the hundredth anniversary of Allessandro Volta's death. In his lecture, Bohr said:[17]

> The very nature of the quantum theory thus forces us to regard the space-time coordination and the claim of causality, the union of which characterized the classical theories, as complementary but exclusive features of the description symbolizing the idealization of observation and definition respectively.

Bohr's idea has spawned a literature in which complementarity is applied to the relations between science and religion, between different religions, and between different aspects of religious traditions. In a widely cited "classic", C. A. Coulson, an Oxford mathematician, defended the view that science and religion are alternative but complementary forms of truth. More recently, Ian Barbour has suggested that complementarity is applicable *within* religious traditions, which are characterized by a shared paradigm, but not to ideas from different traditions, such as the Hindu Brahman and the Christian God.[18]

While some proponents of complementarity parallels have attempted to identify Bohr's views carefully and unequivocally, they have generally been too intent on defending analogies to examine the pitfalls of their efforts. They have little to say about (a) the fact that Bohr himself never defined "complementarity" clearly and unequivocally; (b) Einstein's despair of grasping the principle of complementarity in spite of "much effort", and (c) the exchange between C. F. von Weizsäcker and Bohr, following von Weizsäcker's conclusion in a comprehensive article on complementarity (written in celebration of Bohr's 70th birthday) that he had misinterpreted the principle for twenty-five years; upon asking Bohr whether his revised interpretation was valid, he received a negative reply. Jammer contends that the closest Bohr came to an explicit definition of complementarity was in 1929, when he wrote that the quantum postulate " ... forces us to adopt a new mode of description designated as complementarity in the sense that any given application of classical concepts precludes the simultaneous use of other classical concepts which in a different connection are equally necessary for the elucidation of phenomena".[19] Even critics of the application of complementarity outside of physics sometimes fail to note the ambiguity and vagueness surrounding the principle.[20]

In considering this particular problem in parallelism, it is useful to note Jammer's distinction between the "complementarity interpretation" in quantum physics, and the notion of "complementary pairs" in the history of ideas. Examples of the latter range from Zeno's paradoxes to the medieval idea, "duplex veritas", taught by the Averroists and discussed by Duns Scotus and Siger of Brabant. Jammer points out that the "complementarity interpretation" is distinguished from complementary thought in general by (a) the fact that it is rooted in empirical findings rather than "merely abstract speculation"; (b) the specificity of the idea in terms of the impossibility of reconciling causal analysis and a space-time description; and (c) the link between complementary descriptions and complementary experimental arrangements.

Von Weizsäcker made a distinction between *parallel complementarity* (relevant in the case of position and momentum) and *circular complementarity* (relevant in the case of the Schrödinger function, and the space-time description). Bohr, a circular complementarist according to von Weizsäcker, rejected this distinction in arguing that complementarity is only applicable between phenomena. Jammer's interpretation is supported by Bohr's student, L. Rosenfeld.[21]

In general, the case of complementarity underscores the dangers of generalizing ideas or concepts which (a) have not been rigorously defined within their original realm of application; (b) are, insofar as they are explicitly defined or rigorously conceptualized, specific to the substance, logic, methods, and theories of their original realm; and (c) may appear superficially to have kinship with very general patterns of human thought (for example the complementary pairs notion) but which, because in part of the second point on specificity, may actually represent a different and perhaps a new pattern. The case of complementarity leads to another problem in parallelism, the function of "paradoxes" in physics and mysticism.

PARADOXES

Complementarity parallels have been drawn on to show that science and religion both deal in paradoxes. Austin, for example, explored the possibility of adopting a complementarist approach to paradoxes in religion, such as the idea that the Christian God is both a rigorous judge and a merciful father. He was criticized by physicist James Park who argued that there are no paradoxes in quantum mechanics, and that Austin's use of complementarity as a device for dealing with paradoxes had nothing to do with physics.[22] Nonetheless, the literature of modern physics is filled with references to paradoxes. The term appears in the titles of philosophical as well as technical treatises on physics, in the texts of numerous papers and chapters, and in "problems" such as Schrödinger's "cat paradox" in quantum theory and the "clock paradox" in relativity theory.[23] It has been suggested that quantum physics generates paradoxes because (a) quantum mechanical systems generally fail to obey the distributive law of classical logic; and (b) the elements of Boolean algebra cannot be used to represent the empirical calculus of classes in such systems.[24] Given the widespread discussion of paradoxes in the physics literature, it is easy to see why Capra identifies paradoxes as one of the parallelisms linking physics and mysticism. Granted that paradoxes appear in the literature of physics and mysticism, the question is whether they have the same or different functions in the two fields. The parallelist contention that they have the same functions in physics and mysticism is not readily justified. Paradoxes in mysticism are generally part of the nature of things. In physics, by contrast, they are subject to study with the expectation that they will be "resolved" – that is, brought into the sphere of rational comprehension through the development of new levels of awareness associated with advances in mathematics and physical theory. This orienttion to paradoxes in physics is easily established. D'Espagnat, for example, applauded the Copenhagen interpretation in quantum physics because it emphasized formulating questions in terms of conceptually possible experiments and had as one of its consequences the avoidance of "spurious paradoxes and ambiguities". Bohr, noting that contemporary scientific theory characteristically had to be "sufficiently foolish" to be justified, conceived scientific progress as a process of removing foolishness and paradox from concepts. Similarly, Einstein envisaged scientific progress as a process of running away from "miracles". Kuznetsov assumes that Bohr's "foolishness" and Einstein's "miracles" are roughly equivalent. Terletskii defines paradoxes as unexpected consequences of a theory, and goes on to note that the resolution of paradoxes is a factor in the evolution of theories. It might be argued that there is a parallel between "enlightenment" and Terletskii's conception of the functions of paradox in physics. Mystics who view "enlightenment" as a higher form of comprehension and consciousness than science would probably dismiss such parallels as, at best, spurious.[25]

Mystics seem to be intent on keeping paradoxes intact or using them as devices for enlightenment. This seems to be the case for the koans of the Zen master. Capra stresses this usage in Zen, but at the same time fails to represent the function of paradoxes in physics accurately. He treats the "wave-particle duality" in a way that underscores its "mystery" and obscures the "resolution" of this alleged paradox in modern physics.

The parallelist approach to paradoxes encounters a further and complicating difficulty in the predilection of Hwa-Yen Buddhists for "resolving" Zen-type paradoxes. Chang notes that Zen monks turn to Hwa-Yen for spiritual guidance through the "bewildering" Zen Path and sensible solutions to abstruse Zen problems.[26] Perhaps there is a case here for someone who wants to argue for a paradox parallelism in terms of an orientation to resolution. In any event, there are pitfalls here which reflect the existence of different viewpoints on the naure and functions of paradoxes in physics and mysticism.

ONENESS, TOTALITY, AND INTERPENETRATION

One of the core arguments for parallelism rests on the correlation between (a) the essential teachings of mysticism on oneness or unity, totality, and interpenetration, and (b) the "wholistic" direction of modern physical theories, especially bootstrap physics and monistic geometrodynamics. Capra considers the essence of the Eastern world view to be an awareness of unity in the things and events of everyday experience. The parallel with physics is based on the view that quantum theory has (a) abolished the idea of separated entities; (b) introduced first the observer, then the participant, and most recently consciousness into the heart of physical description and explanation; and (c) stimulated the articulation of a wholistic view of mental and physical relations. The bootstrap idea of "self-consistency", in Capra's view, is not unlike the mystical idea of the unity and interrelation of all things, a parallel he discusses under the subject of "interpenetration".[27]

In Hwa-Yen Buddhism, there are several notions which bear on this particular argument for parallelism.[28] They are (a) mutual entering: all things and principles are mutually dependent and interpenetrating ("mutual immanence"); (b) dependent-arisings: all things exist conditionally and momentarily, and have no "selfhood" or "true being"; this concept simultaneously represents the doctrine of (c) Sunyata, that is, Voidness, or Emptiness: this does not mean "nothingness" or "annihilation", but rather "no-thing-ness"; in the words of Nagarjuna, the founder of Mahayana Buddhism:

> It is because of Emptiness that all things and events can be established. Without Emptiness nothing can be established.

This conception of the Void as the source of all things, as I noted earlier, is correlated by Capra with the notion of "field" in modern quantum physics.

The bases for correlating these ideas in mysticism with the concepts of modern physics are references by Bohm and others to the requirement in quantum theory that the universe be treated as "a single, indivisible unit".[29] One basic question this idea raises concerns whether we are dealing with an updated version of "wholism" in physics, or with a new concept which transcends the cycles of "wholes and parts" approaches in the history of ideas. Max Jammer suggests that we are dealing with something new. He notes that the relational conception of "state" in quantum theory, as initially outlined by David Bohm, is unique because it renounces the classical analytical principle variously stated as "dissecto resolutivo" (Galileo), and Descartes' Second Rule of Investigation. In general relativity theory, a comparable development has occurred with the emergence of an apparent need for a continuum ontology. Graves views this as unique in the history of physical theory. He echoes Bohm on quantum theory and Chew on bootstrap physics when he speculates that modern general relativity – or geometrodynamics – may require that the universe always be treated as a whole. This would be the case if it turns out that no arbitrarily specifiable boundary conditions can be established for general relativity theory. Intuitively, at least, this seems to be akin to the self-consistency requirement in bootstrap physics.[30]

The term geometrodynamics is applied to post-Einstein developments in general relativity in the works of John Wheeler and others. The introduction of the T_{ik} tensor in general relativity theory has broken down the sharp conceptual distinction among mass, energy, momentum, stress, etc. This is considered at some length by Terletskii. He argues that when the energy-mass equations are considered in a three-dimensional noncovariant formulation, inertial mass is conceived as a scalar that varies with velocity. In a strictly relativistic four-dimensional theory, by contrast, there is only the concept of proper mass (M) inextricably linked to momentum and energy. When $E = mc^2$ is considered in a strictly four-dimensional perspective, it must be written as a definition of the mass vector in terms of the energy vectors (that is $E_k = m_k c^2$), rather than a law relating two qualitatively different quantities. From this perspective energy and mass are different projections of the same physical quantity, and all the energy-mass equations (e.g., $E = mc^2$; $m = 1/c\sqrt{(E/c)^2 \ \theta \ (p)^2}$, where p = momentum; etc.) " . . . merely express the organic interrelationship between the concepts of mass, energy, and momentum".[31]

Graves provides more grist for the parallelist mill by conceiving geometrodynamics as a monistic theory in which space-time ("like" ch'i, the Great Void, the quantum field, and the hadron bootstrap) is the source and medium of all interactions. Its parts act and are acted upon by each other; it is a unified whole with global, topological, and local characteristics, and it is the only thing in the universe that is "really real". Graves is not a proponent of a static Parmenidian monism, but one which is internally differentiated to a high degree. In this sense, Graves' monism is different from Spinoza's "God-Substance", and lacks the teleological aspect of Hegelian systems. There is yet more grist in the widely unknown or unappreciated fact among persons with a lay interest in physics that the geometrodynamic interpretation presents general relativity as an already unified field theory. It combines Einstein's equations with the appropriate generally covariant form of Maxwell's equations to give a theory in which gravitation and electromagnetism are represented as aspects of a single tensor field.[32]

The general relativity theorist says that the geometrodynamic curvatures extend throughout space-time, and thus each body or source can be conceived to be everywhere at the same time; the Hwa-Yen Buddhist conceives the universe in terms of "realms-embracing-realms" and the "all-in-one and the one-in-all". The general relativity theorist views boundary conditions as arising from the content of space-time, giving us one self-contained process; the Hwa-Yen Buddhist enters Samadhi – the dynamic state of Enlightenment – and is liberated from all "obstructions" or "boundaries".[33] Note, once again, the congruence between the geometrodynamic interpretation of general relativity theory and the bootstrap hypothesis.

The gist of all this from the physical side is a conviction among physicists that they are being pressed toward new ways of thinking about systems of "many interconnected components".[34] Parallelists see this as a movement in the direction of mystical conceptions of reality.

The fundamental pitfall encountered in this instance of parallelism can be considered "thematic", in Holton's sense.[35] Holton conceives themata as " . . . fundamental preconceptions of a stable and widely diffused kind that are not resolvable into or derivable from observation and analytic ratiocination". Thema are invariably associated with antithema. Science is conceived to be a dialectic process which, temporary victories for one themata or another notwithstanding, opposes thema and antithema and "energizes" research. "Wholes" and "parts" can be conceived as thema and antithema (or vice versa), each rising and falling with respect to the other as the history of ideas in general, and of science in particular, unfolds. The thematic pitfall involves the problem of whether "wholism" in modern physics is similar or otherwise analogous to "wholism" in mysticism, and to "wholism" in pre-modern physics. Gale, for example, has argued that there are many similarities – or parallels – between Chew's bootstrap ideas and Leibniz's conception of monads.[36] These are the same sorts of parallels Capra and other parallelists point out in comparing bootstrap with Buddhism, as well as with Leibnizian physics. If we are guided by a Holtonian thematic perspective, we should not be surprised to find such parallels. It remains to establish whether we are dealing simply with long-established themata, or whether we are dealing with new thema reflecting new orders of human thinking.

This pitfall is perhaps nothing more than the tendency to try to understand the present in terms of the past, the unknown in terms of the known, a tendency manifested in the ease with which we draw analogies within and across all systems of thought. For the case at hand, one basic reason for questioning these analogies lies in the mathematical requirement of nonlinearity in bootstrap physics and geometrodynamics, a requirement whose uniqueness suggests that we are not dealing with just another cycle of "wholism" in modern physics. The conceptual significance of this is that with linearity, sources and effects are independent and can be added. In the case of geometrodynamics this means that if we are given a total field, we can analyze it uniquely into parts, and assign each part a source. In the non-linear situation, we could not expect to break up the field into a set of independent parts. It would have to be treated as a whole. D'Espagnat, interestingly, refers to "physical reality" as everything that strictly obeys the linear laws of quantum mechanics, a situation that holds as long as "consciousness" does not enter the picture. It is important to note the conjunction between nonlinearity requirements and concerns about consciousness in the works of Chew, Wigner, and others.[37]

Bohm is the most explicit among "new physics" advocates in suggesting that we are not dealing with a wholism cycle in quantum and relativity theory.[38] Thinking in terms of wholes and parts and the classical dynamic of analysis and synthesis is no longer valid in Bohm's view of "holonomy", a process of theoretical insight in which "new wholes" are continually emerging. Bohm describes this distinction in terms of the analogical contrast between the lens and the hologram. The lens brought object and image into sharp relief, and "strengthened man's awareness of the various parts of the object and of the relationships between these parts"; it furthered classical analytic and synthetic thinking. He argues that, by contrast, in contemporary physics " . . . an instrument tends to be relevant to a whole structure, in a way rather similar to what happens in a hologram":

> There is the germ of a new notion of order here. This order is not to be understood solely in terms of a regular arrangement of objects (e.g., in rows) or as a regular arrangement of events (e.g., in a series). Rather, a total order is contained, in some implicit sense, in each region of space and time.

Whether in fact nonlinearity requirements, the integration of consciousness into the laws of physics, and the "holonomy" idea represent (a) a temporary orientation based on the thematic cycle of wholes and parts, or syntheses and analyses, or (b) a true break with or transcendence of that cycle is by no means established. The suggestion that modern physical "wholism" is unique, however, may vitiate any parallels with thematic wholism in mysticism. On the other hand, while mathematical nonlinearity is not a factor in mystical experience, it certainly would not do to assert dogmatically that mystical "wholism" is a standard thema rather than an experience of a quite different kind, one that physicists may be converging on.

KNOWING AND CONSCIOUSNESS

There are two views of knowing associated with parallelism. The "conservative" view is that science and mysticism are independent but complementary ways of knowing; they represent, respectively, rationality, and intuition. This viewpoint is reflected in many of the works on complementarity in science and religion referred to earlier. The "radical" view is that there are three modes of knowing. Siu, for example, distinguishes rational, intuitive, and no-knowledge modes of knowing.[39] No-knowledge, according to Siu, is experienced beyond the limits of rational and intuitive modes of knowing. The realm of no-knowledge is devoid of shape and time, and transcends events and qualities. It is the realm of the silent apprehension, of the "undifferentiable whole" that Capra seems to have in mind when he discusses mysticism, but which he equates with intuition.

A more dynamic version of the radical view of knowing is suggested by TenHouten and Kaplan.[40] They view science and mysticism as dialectically interrelated complementarities. This view is rooted in the idea that one hemisphere of the brain functions primarily in an appositional mode, and the other hemisphere in a primarily propositional mode. Comprehensive consciousness and knowing, including rational, intuitive, and no-knowledge dimensions, may involve the dialectical interaction of the two hemisphere-specific modes, and may have its physiological basis in the corpus callosum that links the two halves of the brain. These ideas are highly speculative. The "modes" pitfall rests on the fact that we do not have a convincing theory of knowing. Are there two modes of knowing, three modes, or more? Are modes of knowing an artifact of analysis rather than a fact of consciousness? If there are different modes of knowing, how are they interrelated? What is the relationship between modes of knowing and the structure and function of the brain? The TenHouten and Kaplan theory of inquiry, based on studies of the hemispheric functions of the brain, generally supports parallelism. Their interpretation of so-called "split-brain" research, however, must be scrutinized critically given the lack of a fully-developed and coherent theory of brain functions.[41]

A related issue is the extent to which modes of knowing are inextricably linked with particular relationships between human beings and their surroundings. Needleman, for example, notes that the fact that some scientists meditate or practice yoga is neither an indication of harmony between physics and mysticism, nor a bridge to the "quiet attempt to observe the mind" some people assume it to be. This viewpoint is echoed fervently and eloquently by Roszak, who argues that alienation and cultural blight are rooted in the "objectified worldview of natural science" – that is, the physics of Newton, Einstein, and Bohr, and the biology of Darwin, Crick, and Watson.[42] This suggests that science can only progress through manipulation of the material environment, and that this is fundamentally opposed to and irreconcilable with the religio-mystical approach to "appreciating" reality. However, science may be moving in a direction that will force us to reassess the degree to which we rely on experimentation to determine the "truth value" of hypotheses. We may be approaching a time when it will not appear incredible to consider the possibility that "truth value" can be determined by a state of consciousness independent of the direct physical manipulation of variables in experiments. This notion has been adumbrated in contemporary writings on the philosophy of science such as Holton's critique of experimenticism, and Hooker's discussion of non-linguistic dimensions in inquiry.[43] Ravindra's position on this question is exemplary:[44]

> . . . whatever we can study from the scientific point of view of manipulation and control – whether it is universe, man or divinity – has been produced, at least partly, according to our plans; it is something that can be compelled by us to yield answers to our questions. It cannot be higher than us; for that which is higher – in nature no less than in man – can neither be coerced nor violated by us. We can prepare ourselves for it and wait, actively making an effort of attention, observing without violence.

This position will be rejected by physicists and scientists in general insofar as it expresses a commitment to noumena – ultimately unknowable things-in-themselves. Mystics and physicists may agree that ordinary sensory experiences must be transcended in order to comprehend reality. They must part company, it would appear, when and to the extent that the mystic remains "constrained" in quiet appreciation while the physicist presses forward, transcending ordinary experiences and at the same time actively seeking to explain what he/she experiences. Any consideration of possible convergences between physics and mysticism must deal with whether, how, and to what extent some version of realism links these two modes of knowing.[45] Capra's view that an "empiricism parallel" links physics and mysticism can be considered in the light of this discussion.

Capra claims that physics and mysticism are both strongly observational.

This is not self-evident, even though D'Espagnat, for one, supports Capra's view of the centrality of observation in modern physics.[46] The mystic's distinction between (a) seeing, looking, and watching, and (b) thinking, may be a semantic trap. The meditative state might be better described as a state of non-ordinary thinking than as an observational state. The centrality of mathematical thought in physics (consider Einstein's work, or the role of group theory in elementary particle physics) could be pointed to in support of the argument that theory rather than experiment, or abstract reasoning rather than observation, is the essence of modern physics. If there is any parallel here at all, it might more likely lie in the direction of abstract thought rather than empiricism. It may be that what we are dealing with here is a manifestation of different ways of attending to things, an idea introduced earlier in the section on space-time parallelism. If this is true, and altered states of consciousness are based on changes in how we attend to things, it may be possible to correlate states of consciousness with criteria for determining "truth value", and furthermore to show that these criteria are interrelated and interdependent. This would help to resolve the question of the relationship between physics and mysticism in terms of modes of knowing and consciousness.

The novel and speculative nature of these ideas makes it impossible to rely on them in evaluating parallelism. They are pitfalls which parallelists and their critics must consider and weigh in establishing their positions.

ANTICIPATION OR CONVERGENCE?

There are two pitfalls of interpretation in parallelism. One is associated with the assertion that the parallels manifest anticipations – that is, modern physics was anticipated by ancient and more recent mystics. The second is associated with the idea that modern physics is converging towards mysticism. The anticipations thesis looks less and less reasonable the more deeply we examine the parallels. It reflects weak analogies at best; and these analogies are possible only because there are certain basic and recurring themes in human thought.[47] One explanation for the apparent convergence of modern physics and mysticism is that as physicists have probed deeper and deeper into nature, they have been forced to abandon more and more of the images and concepts of ordinary experience. Mystics, however, have been probing non-ordinary reality for centuries. If it is assumed that the non-ordinary reality open to mystics is in the same realm as the non-ordinary reality open to physicists, it should not be surprising to find that the verbal descriptions of reality by physicists and mystics exhibit correspondences. It is conceivable that such correspondences are implied in Bohm's speculation that thought processes are critically dependent on quantum-mechanical brain elements, and that they provide direct experience of quantum theory in the way that muscular forces provided direct experience for classical theory.[48]

An alternative explanation is that physics is passing through a stage in which the boundaries of experiences that we have learned to express in common languages are being penetrated, and these experiences are pressures for creating a new, relevant vocabulary. This is occurring on the level of mathematical formalism and physical theory, and on the level of common language. If attempts to describe physical reality sound like mysticism, it may be because there are certain general linguistic patterns that people turn to when they have to describe the "indescribable". If this is the case, it is unlikely that physicists will become mystics. As their inquires proceed, and their experiences become more ordinary for them and for the general public, the common language can be expected to change and the parallels with mysticism will disappear.[49]

Paradoxes in physics and mysticism may arise at the intersection of the processes of symbolizing and experiencing. Sooner or later, if we are pressing our inquiries continuously forward, we will be led beyond the limits of our language. This process will approach some sort of limit in mysticism where the goal is a certain state of consciousness. But in physics, ideally, new experiences are assessed and internalized, old symbols are transformed, and new symbols are created. What appears to be a convergence between physics and mysticism may simply reflect the process of transforming and creating symbols, which will lead the physicist on to new intellectual paths that diverge from the current apparent parallels, analogies, and convergences with mysticism; and this seems to have happened, in fact, in the case I mentioned above of the existence proof for plural structures of four-dimensional space-time.

Physicists, according to Capra, "have made a great step towards the world view of the Eastern mystics".[50] The pitfalls I have discussed illustrate that this claim cannot be readily validated. Even if Capra is, in some sense, correct, it is not intuitively obvious nor logically necessary that the next step (or steps) must be in this same direction. We must also entertain the possibility that the "great step" was, from the point of view of the future of physics, a step in the wrong direction.

IDEOLOGY

There is, finally, the pitfall of ideology. Capra, for example, refers to the "marked anti-scientific attitude" of people who are attracted to Eastern mysticism. One of his objectives in *The Tao of Physics* is to improve the image of science by showing the essential harmony between modern physics and Eastern mysticism. There is no need for seekers of wisdom and spirituality, Capra argues, to abandon physics; it too can be a "path with a heart" leading to self-realization and spiritual knowledge.[51] Parallelism can, therefore, function as a defensive justification for and explanation of the scientific approach, and a device for improving and supporting the image of science and scientists. Similarly, advocates of religious "truth" have responded to the threat posed by the successes of scientific "truth" by seizing on alleged parallels between religion and science to justify and explain their ways of thinking and behaving. This is illustrated in the literature on complementarity in science and religion I discussed earlier. I will treat this problem in terms of an "interests" thesis in Section III.

PHYSICS, MYSTICISM AND SCIENTIFIC CHANGE

The pitfalls I have identified indicate that the parallels between modern physics and Eastern mysticism may be spurious for reasons ranging from semantics to ideology. The fact that it is possible to identify parallels, analogies, and convergences between modern physics and mysticism may reflect a temporary condition in modern physical theory. Physics, and science in general, can be expected to develop and change in fundamental ways under the imperative of pressing inquiry ever forward. Mysticism, by contrast, appears to have arrived at certain "ultimate" experiences and truths, at least within the boundaries of current human biological potentials, under a soteriological imperative that subordinates, ignores, and subverts active, open-ended inquiry. This does not mean that parallelism must be firmly and finally rejected as a manifestation of substantive linkages between physics and mysticism. In concluding this chapter, I wish to explore some reasons for moderating the skepticism prompted by considering the pitfalls of parallelism.

The TenHouten–Kaplan outlook on inquiry suggests that modern physicists, more so than their predecessors, may be working on or toward the level of "transcendental", or "higher" brain functions. Mathematical formalism may, at the frontiers of modern physical theory, involve mental functions different from those associated with the capacities of either brain hemisphere, or with the "normal" capacities of the brain revealed in ordinary experiences. Mystics may be tapping the same higher level functions. It may be that such experiences have the potential for revealing the "deep structure" of the universe. This would follow from the hypothesis that the brain as a whole and the universe as a whole are related isomorphically. The implication of this hypothesis is that the greater the degree to which the full potential of the brain is actualized, the greater will be the degree to which the "deep structure" of the universe will be revealed. This idea is not unrelated to the bootstrap notion that consciousness is embedded in the system of self-consistency we understand as the universe. A "limited isomorphism" would probably be more consistent with an evolutionary viewpoint in the sense that the human brain is not (and perhaps cannot be) in total and perfect correspondence with a universe that is evolving, and very likely in advance of the capacity of consciousness to keep up with it. Isomorphism, even in the "limited" sense, coupled with the operation of physicists and mystics on the level of higher brain functions, would help to explain and support parallelism. The substance of these speculations would hold even if, instead of being close to tapping the full potential of the brain, we were witnessing a "leap" to a broader and deeper comprehension that falls short of what is ultimately possible. The best way to characterize what is happening to our capacity for comprehension from the speculative perspective I have sketched may be "relationalism".

Parallelism may be a direct or indirect manifestation of the emergence of a relational way of viewing reality. Relationalism may be either (a) a manifestation of some fundamental themata (in the Holtonian sense), perhaps "wholism", that has been at times central and at times peripheral in the history of ideas, or (b) a new themata, as suggested in my earlier discussion of non-linearity and Bohm's notion of "holonomy". In addition to being a feature of parallelism, relationalism appears in (a) geometrodynamic monism, (b) bootstrap interpenetration, (c) Helier Robinson's metaphysics, and (d) L. L. Whyte's universalistic world view. I have already discussed parallelism, geometrodynamic monism, and bootstrap interpenetration. I want to add a brief note on relational metaphysics and the relational world view.

Helier Robinson has proposed a relational metaphysics which, he argues, reflects and is consistent with contemporary physical science. His system turns out to be a bootstrap one. The universe is conceived as a "singular possibility"; it is the best of all possible worlds in the sense (recall Chew) that it is the only world possible. Reality is a "single polyadic relation". Robinson does not refer to parallelism, nor does his work suggest ways of avoiding the pitfalls I have pointed out. The fact, however, that he arrives at ideas consistent with parallelism and based on relationalism at least strengthens the imperative for critical studies of parallelism, and reinforces the notion of an emergent relational view of reality.[52]

L. L. Whyte has stressed precisely the relational hypothesis I am considering. He conceives a change in our view of reality which will emphasize "hierarchies of morphic process", and the human being as a "coordinated organism". His view of healthy, joyful human beings "perpetually creating new unities" echoes Bohm's conception of holonomy. He also views the emerging unification of world views as one of "circular self consistency", a notion in line with Chew's self-consistency hypothesis. Whyte argues that twentieth-century scientific and philosophical thought has moved toward, but not yet culminated in, a truly relational theory. Such a theory would mean, he writes:[53]

> . . . that experienced qualities and measured quantities could both be derived from a single basis, particular qualities being experienced and quantities measured in certain contrasted circumstances . . .

Changes in spatial relations, Whyte argues, can account for both subjectively experienced qualities and objectively measured quantities. Relationalism implies an alteration in our view of what is "real", and of "existence". Whyte conceives this as a movement which will eliminate any remaining vestiges of classical mechanism. This suggests what in Kuhn's terms might represent a "paradigm shift", or in Holton's terms the dialectic outcome of a cyclic struggle between thema and antithema.[54] This would be a significant development, signaling a "scientific revolution"; but then it is possible that relationalism portends a *second* scientific revolution. Such a revolution would establish a new outlook on the nature of scientific inquiry (beyond a "paradigm shift" that occurs within the boundaries of the scientific revolution associated with Newton, Galileo, *et al.*), and affect and integrate a broad, if not the full, range of modes of human inquiry.

We must, finally, consider the possibility that there is a social factor operating in parallelism. This is an obvious imperative of the sociology of knowledge, but one that still lacks the strong foundation of a fully-articulated theory. We have "paradigms" such as Paul Forman's case study of the link between quantum theory and Weimar culture, and Lewis Feuer's weaker analysis of the origins of Einstein's theories in "social relativity".[55] Is it possible that parallelism and relationalism reflect and generalize our increasing awareness of sociological and ecological phenomena? Are these causally related, and if so which is cause and which effect? Are these phenomena simultaneous effects of some underlying causal factor, perhaps a neurological one? Or are they outside the realm of classical cause and effect, self-exemplifying the emergence and development of relationalism? If a social force is indeed operative, is it one rooted in the evolution of our understanding of reality (for example, sociology), or one that reflects a more immediate need to offset the disastrous human and environmental effects of unfettered individualism, and specialization without interdependence? Obviously "social factor" can involve all of these things, as well as be a part of a general trend toward relational theory and a relational perspective.

I have been concerned in this Section with sketching the parallelist argument and identifying pitfalls in the search for parallels between physics and mysticism. In concluding the present chapter, I have speculated that even if the assumptions of parallelism are – from a critical viewpoint – untenable, the *fact* of parallelism may be related to changes in science and society. In Section III, I will return to this problem and examine it from a sociological perspective. But first, I want to outline a sociological way of looking at physics and mysticism. I view mysticism and physics as *social facts*, whereas parallelists tend to treat mysticism as a "mystery" and physics as a set of facts about reality accessible to laypersons as well as specialists. As *social facts*, mysticism and physics represent *adaptations* in sociocultural milieux. In Section III, I will introduce an "interests" thesis, and discuss what parallelists *gain* through *their* particular treatment of physical and mystical ideas. In Section IV, I will discuss alternatives to the parallelist adaptation or "strategy".

In the Appendix to this chapter, I examine Joseph Needham's views on Chinese and modern science. Needham does not make crude parallelist claims about ancient Chinese sages anticipating Einstein. But he does offer a form of parallelist argument. Like Capra and other parallelists, he believes we are living in troubled times; and like the parallelists, he looks backward in time for a useful strategy – in this case, it is the organicist world view of Chinese science. Many of the issues raised by Needham's methods and arguments are relevant to a critical understanding of parallelism.

APPENDIX B

THE SEARCH FOR CHINA'S ARISTOTLES, GALILEOS, AND EINSTEINS

In all of his writing prior to the "author's note" that introduces Part II of Volume 5 (in which some tentative qualifications which I will discuss shortly are introduced), Joseph Needham describes the Chinese scientific tradition as pre-, proto-, or para-scientific.[56] His point of reference is *modern* science, which he conceives to be the latest stage in the evolution of science. In contrast to the two earlier stages, *primitive* and *medieval* science, modern science does not have an "ethnic stamp". Thus Needham refers to modern science as it emerged in the West's Scientific Revolution in phrases such as "the universality of modern science", "science in the fullest sense", "perfected world-view of natural science", and "universally valid world science". In the Scientific Revolution, according to Needham, "the basic technique of discovery" and "the full method of scientific investigation" were discovered, the "absolute universality of mathematics" was established, and the conditions for generating "a body of incontestable scientific truth acceptable to all men everywhere" based on the "universal language of mathematized hypotheses" set up. Universalistic science, potentially open and comprehensible to all human beings, is viewed as the foundation of a new universalism which is preparing the way for the unification of "the working peoples of all races in a community both catholic and cooperative".

We are, for the moment, Needham believes, in troubled times. This is not, in his view, because of modern science as some critics of science contend. Needham considers science "neutral". The fact that good and evil effects attend the development of natural knowledge is, he argues, a fact of human nature. This orientation has as one of its consequences (explicitly expressed by Needham) the fact that in some cases at least scientists must engage in non-benevolent acts in order to insure long-term benevolent results in the pursuit of knowledge: this is a *sine qua non* of scientific method. This viewpoint is developed further in Needham's criticism of Said Hussain Nasr's argument that modern science never arose in Chinese or Islamic civilization because of metaphysical doctrines and traditional religious structures that prevented profaning nature. Needham argues that science demands working *as if* Nature is "profane". In part, Needham wants to point out that we cannot do science if we are going to divide the world up into things that *can* be studied scientifically (the profane) and things that cannot (the sacred). He succeeds, however, in lending support to three positions – the neutrality of science, the profanity of nature, and the need for non-benevolent acts – that appear to be inconsistent with his socialistic sympathies.

There is some confusion in Needham's work because of the fuzziness of the concepts "modern science" and "Scientific Revolution" in his work and in the history of science in general. Needham relates Chinese science to three versions of modern science: (a) Galilean-Newtonian science (the mechanistic version), (b) Einsteinian-Planckian science (the field, organicist version), and (c) future science (fully and unequivocally organicist). Thus, when he relates Chinese science to version (a), Chinese science appears to be very different in its philosophical foundations, but pre-, proto-, or para-scientific in terms of scientific attitudes, theoretical orientations, and methods. When compared to relativistic-quantum mechanical science, China's organicist perennial philosophy fares well as an anticipation of the "modern" scientific world view. Since future science will, in Needham's view, probably be even more organicist than relativistic-quantum mechanical science, China's scientific achievements are most notable in this third comparison.

Following this line of inquiry, it is interesting to note that the "universal science" introduced by the Scientific Revolution Needham eponymizes as Galilean lasts, in one of his accounts, for hardly two centuries. In a matter of 163 years, from the birth of Galileo to the death of Newton, Needham takes us from a Revolution to the seeds of the destruction of the new science introduced by that Revolution, sown by Leibniz. Leibniz is presented as the leading figure in the movement to replace the mechanistic science of Galileo with a more organicist science. (Leibniz is supposed to have been influenced by Chinese ideas, but on arguable evidence). This confuses the issue of what modern science is and what the Scientific Revolution accomplished. It reflects the lack of careful conceptualization, and the inevitable difficulties that arise when "heroes" such as Aristotle and Galileo are relied on to explain social facts. Aristotle and Galileo are the two central figures in one of Needham's most confusing exercises in the history of science, the "parallel heroes" question.

Did China produce an Aristotle or not; and if China did not, was this a hindrance to the development of Chinese science on the path to modern science? For Needham, at times, and for some historians of science, Aristotle is an important link in the chain of activities and processes that leads from pre- to modern science. Needham cites Aristotle, for example, as a figure in the trend to experimentation which started with him and "led to Galileo through Leonardo". Aristotle is also portrayed as the pre-scientific polar genius on the theory of impetus, with Nicolas Oresmus the medieval transitional figure, and Newton at the modern science pole.

There are dangers in seeking parallel heroes, that is, looking for China's Aristotles, Galileos, and Newtons. Such an approach imposes specific historical biographies on other people and obscures their individuality, and the ways in which they reflect and reflect on their particular cultural settings. It also expresses a unitary view of science and scientific contributions. Even if there is some heuristic value in this kind of exercise, it is hindered by Needham's equivocations. Thus, he suggests in one place that China produced no Aristotle, and at the same time points out that Aristotle and Aristotelian logic were necessary ingredients of modern science. Needham cannot, however, unequivocally assert that the term *fa* used by the Mohists did not include all of the Aristotelian causes, nor can he maintain without uncertainty that the Mohists had not in fact developed Aristotelian logic, or that they were not capable of inventing it, or furthermore that Aristotelian logic does not exist "informally" in Chinese thought. In addition, Needham finds discussions in the Chinese literature that are, in his view, as sound from a scientific viewpoint as anything in Aristotle. The discussions in the *Pao Pu Tzu* (Book of the Preservation-of-Solidarity Master), written in the early fourteenth century, are exemplary. Furthermore, Needham notes that scholars from Francis Bacon to A. N. Whitehead considered Aristotelian logic more of a hindrance than a help in the emergence of modern science. Given these problems, it does not appear that the comparative history of science in China and Western Europe can gain much from a search for China's Aristotle.

For Needham, Galileo is the focal point of the Scientific Revolution, the central figure in the mathematization of science, and the developer, in "almost perfect form", of the experimental-mathematical method. This view of Galileo as the hero of the Scientific Revolution may, however, create confusion, especially when Needham takes us hunting for China's Galileo. Needham points out, in the first place, that Galileo had "precursors" (e.g., Buridan, Bradwardine, Philoponus, and Nicolas Oresmus). In addition, Needham argues (weakly) that the theorizing and experimenting components of the Galilean method were possessed by, respectively, the scholastics and the higher artisanate in pre-Galilean Europe.

There are at least three points to consider in weighing the place of Galileo in the Scientific Revolution, and the significance of Needham's contention that China did not produce a Galileo. First, historians of science have not abandoned Galileo as a hero; but increasingly, they have made him share the pedestal with other heroes, notably Copernicus, Kepler, and Newton. Second, historians of science are not unanimous and unequivocal about Galileo's place in modern science. It has been suggested, for example, that what we see as "new" in Galileo's *nuova scienza* depends on what we think characterizes modern science, and in particular modern physics. This includes consideration of the extent to which the mathematization of physics is more an ideal than an accomplishment in Galileo's work. Finally, given Needham's conception of modern science, we can ask whether Galileo is an appropriate representative of that science given his actual methods.

In his introductory note for Part II of Volume 5 in the *Science and Civilization in China* series, Needham acknowledges Nathan Sivin's argument that Chinese science must be seen as developing out of one theoretical state into another, and not as an aborted stage on the road to modern science. Needham admits that traditional Chinese science should not be viewed as simply "a failed prototype of modern science". This is a viewpoint he has defended in the past. It is illustrated by his contention that Taoism retained "unborn within itself, science in the fullest sense", and was oriented in a direction that would have ultimately led to modern science. He admits further that modern science is not the last word, and that everything in the past should not be judged in the light of modern science, nor related to it as a final court of appeal. However, there is some indication that Needham is not fully persuaded by this, and that in fact he misses the point of the criticisms he addresses. The above admissions are immediately qualified: modern science is not the last word, and we must remember its "transitory nature", but it is a "reliable measuring stick"; science today is not "all that science will ever be", but we cannot deny "the fundamental continuity and universality of all science". To the extent that he is persuaded by critics of the positivism characteristic of his earlier contributions, it seems to be in terms of envisioning potentials for scientific change that will allow him to assign an even more exalted position to Chinese science in the history of "world science" (which must now be considered a future development rather than a present reality).

Needham acknowledges that differences concerning how modern science emerged and what its future is likely to be can be entertained. He also acknowledges two points of view on the relationship between Chinese and modern science. One is the viewpoint he still seems to cling to, that the Chinese and Western scientific traditions were on paths directed toward modern science. The second, advocated by Nathan Sivin, is that Chinese and Western science might have been, and still be, on separate paths, with their true merger conceivable only in the future.

Needham convincingly argues that he has brought an unusual combination of talents to bear on the documents and archeological evidence of ancient and medieval China. In particular, he has stressed his education in science as necessary preparation for dealing with the Chinese scientific and technological tradition. This should not be accepted uncritically as a blessing. There is reason to be sceptical of Needham's translations and interpretations. He has an understandably strong – some critics would say "zealous" – commitment to setting the record straight on Chinese achievements in science and technology, long unknown and ignored by Western scholars. He also developed strong intellectual attachments to Whitehead's metaphysics, the Marxian idea of inevitable stages in social evolution, dialectical materialism, and the universality of the modern scientific orientation before undertaking his study of science and technology in Chinese civilization. Given the widely acknowledged difficulties of translating in general, and of translating Chinese in particular, and the possibility that Needham's linguistic skills are not up to the highest standards, it would be foolhardy in the light of Needham's intellectual biography to uncritically accept his continuing discoveries of Whiteheadianism, dialectical thinking, and anticipatory pre-modern scientific attitudes, concepts, and methods in Chinese texts. These discoveries are even more problematic given the fact that Needham's project is not an in-depth, definitive study but, in his own words, "a reconnaissance".

Warnings and cautions, tempered and otherwise, have been duly sounded. William McNeill, for example, has labeled "nonsense" Needham's claim that the Taoists had the idea of "natural selection", and generalized this by implication to all of Needham's efforts to discover parallel ideas in the light of the formidable difficulties of translation. Sivin has argued that we will be bound to trivialities if we forget that Chinese concepts and attitudes – for all the similarity they may exhibit to concepts and attitudes in science today and in the future – were embedded in disciplines "drastically different" from our own in aim, approach, and organization.

One of the things that is missing in Needham's conception of his project and in the comments of his detractors and appreciators is a sense of the comparative history of science as an exercise in the cross-cultural anthropology of knowledge. Let me begin by noting a specific problem in Needham's methodology which suggests the need for an anthropological approach to the comparative history of science. Needham's stress on distinguishing between empirical principles (characteristic of pre-, proto-, and para-scientific activities in China and elsewhere) and systematic theory (characteristic of modern science), and his reliance on a unilinear theory of the evolution of science through three stages – primitive, medieval, and modern – tends to obscure the theory-laden nature of all human observations and activities. Furthermore, the stress on the distinction between empirical and systematic science, and the unilinear theory of scientific evolution insures that we will only be able to understand pre-modern theoretical statements, or infer theoretical understanding by pre-modern people, in terms dictated by modern science. Some Chinese theories might even be overlooked because they are not "modern" in approach. We are also likely to view pre-modern theories as parochial and modern theories as universal without being able to show clearly and unequivocally why modern theories are not also in some sense parochial. Finally, we are likely to conclude that since the theories we read or infer are "incorrect", it is amazing how often, with what skill, and on what scale great technological achievements can occur; two or three thousand years from now, perhaps, our own technological achievements may appear, from a comparable perspective, miracles of "wrong reasoning".

The pre-modern Chinese, so far as we can gather from Needham and other students of Chinese science, were no more or less theorizers than the people of other pre-modern civilizations. Their pre-modern theories were not quantified, tended to be more explicitly integrated into a world view, were independent of experimental evidence in the modern laboratory sense, and were not characterized by the continuity associated with science from the fifteenth century on. The question is: is there more to scientific understanding than modern-like theories that can be written down? What about non-linguistic forms of inquiry and understanding? Most scholars would probably accept the idea that non-linguistic forms of knowledge are part of human thought, but what about the possibility that they may be, in some respects, more important than linguistic forms? The more conventional issue here is the controversy over the relationship between theory and observation. For some philosophers and other students of science, Feyerabend's, Hanson's, Kuhn's and Toulmin's positions on scientific changes as changes in observationals are too strong and destroy the possibility of referring theories to experience. The underlying problems here may, to some extent, be related to Needham's treatment of the relationship between theory and observation. His approach invites critical anthropological scrutiny.

The "psychic unity of mankind" hypothesis or assumption is central to Needham's heart and method, and not unrelated to Sivin's view that we base comparisons on an understanding of Chinese science in its own terms. This idea, which has had a troubled history in anthropology, states that human beings everywhere and at all times (or, at least, over a period during which we can be relatively certain that no major biological changes – especially involving the brain and nervous system – have occurred which would affect mental potential or functions, say the last 35 000 years) have inherited similar propensities and abilities. As a general proposition, and as an egalitarian and anti-racist proposition, sentiment, or metaphor it can, in my view, stand unchallenged. There is, however, a chance that if it is accepted uncritically and axiomatically as a statement about *cultural* potentialities it might obscure or close access to *learned* differences which seem to contradict psychic unity, and incline students in cross-cultural and comparative studies to minimize or ignore the difficulties of understanding a second culture in its own terms.

Consider, for example, Donald Campbell's studies of responses to optical illusions. In undertaking these studies, he noted that if different response patterns were found in different cultures, the first place to look for an explanation should be differences in visual environments. He proposed "the carpentered-world hypothesis": the proportion of artifacts that are rectangular (i.e., the degree to which the visual environment is carpentered) varies across cultures; the greater the degree of carpentering, the greater will be the tendency for people to habitually interpret obtuse and acute angles as rectangular surfaces extended in space. Similarly, the habit of interpreting elliptical retinal images as circles extended in the third dimension might be absent in cultures where the visual environment included objects which were actually elliptical in cross-section.

Campbell also identified the two-dimensional representation of three-dimensional objects as a factor related to illusion responses. Perspective drawing, he pointed out, is a pervasive feature of Euroamerican culture from childhood on, and the techniques or conventions of perspective drawing may be related to the habits of inference which some illusions illustrate. I will refer shortly to the relevance of China's reliance on "parallel perspective" for Campbell's hypothesis.

Campbell's experiments indicate that there are differences in the ways in which people respond to standard (for Westerners) optical illusions. This result could lead to the conclusion that different people perceive and think in radically different ways. The differences Campbell observed, however, were small; and it was because the differences were small that they could be determined. This suggests an *optimistic epistemological hypothesis*: there are differences in perceptual and thought patterns across cultures, but they are small enough to permit cross-cultural understanding. There is also, however, a *pessimistic epistemological hypothesis* implied in Campbell's results. If, Campbell writes, we ask how people discriminate between ignorance, truth, and falsehood, we can hope to answer this question within certain limits when such discriminations are actually made. Such knowledge, however, may turn out to be possible only under rare conditions, and possibly only under experimental laboratory conditions. Campbell concludes that the achievement of such conditions may be entirely lacking in the cross-cultural study of philosophy, and by implication in the study of knowledge and science. We should, he urges, strive to achieve an "insider's view", but recognize (a) the impossibility of achieving more than an approximation, and (b) the fact that a brain which has mapped one environment is a biased machine for mapping other environments.

Campbell proposes the method of triangulation for approaching the limits of certainty in cross-cultural studies. Basically, this amounts to relying on ethnographers *from* plural cultures *studying* plural cultures. It would be enough, he argues, to have two ethnographer cultures focused on one target culture. This, in fact, is what Needham and his collaborators have tried to achieve. However, Campbell is concerned with studying cultures in the present, and his final comment on these matters points out the special (and ultimately perhaps insurmountable) problem faced by students of past cultures. Any given cultural item (including a linguistic item, I would emphasize) must be examined in its total cultural context. Obviously, the student of past cultures cannot realize this goal in the same way and to the same extent that the student of present cultures can. The fundamental problem, however, of special concern for the student of past cultures, is that the context dependence of cultural items "removes the possibility of certainty and makes ever present the possibility of erroneously alleging cultural differences as a result of mismatching". This problem is obviously also present in establishing similarities.

The preceding discussion falls, in general, within the province of the so-called "culture-and-cognition paradox" which I want to consider briefly. The culture-and-cognition paradox, according to Paredes and Hepburn, is this: on the one hand, there is the psychic unity of mankind hypothesis or assumption; on the other hand, anthropologists have shown that cognitions are dependent on environment and culture. The psychic unity assumption leads us to expect that human beings in different cultures will "think" similarly; "paradoxically", we find evidence to the contrary. It is not clear that this is in fact a paradox. In any case, it is considered a paradox by some anthropologists, and is based in part on the attempts by ethnographers to describe and explain different patterns of thought encountered in cross-cultural studies. This has led to a distinction between two basic modes of thought and knowledge, variously referred to as lineal and non-lineal, abstract and concrete, and analytical and relational. In the case of China, a similar distinction has been applied. Chinese thought has been referred to as "co-ordinative" or "associative" and contrasted with the "subordinative" nature of Western thought. Sinologists not only note that the Chinese form is distinct in its emphasis on pattern and organicism, but also that it has its own form of causality and logic.

It might occur to someone familiar with research on the hemispheric functions of the brain that the culture-and-cognition paradox might have some relationship to patterns of hemispheric dominance. To put it simplistically, were the Chinese right-brained and the West Europeans left-brained (following the usual association of the left hemisphere with propositional thought and the right hemisphere with appositional thought)? Paredes and Hepburn have proposed just such an hypothesis to resolve the culture-and-cognition paradox. Their proposal can be summarized as follows: (a) there are plural cognitive processes, but different matchings cross-culturally between particular processes and particular problems; (b) the existence of neurological bases for plural cognitive processes eliminates the need to explain alterations in brain functions from one situation to another in terms of environment; (c) modes of thought appear to be hemispherically localized, but the hemispheres are interconnected and intercommunicate; (d) cognitive strategies are rooted in combinations of mental processes which can be theoretically represented "as an indeterminate number of outcomes of cerebral interaction, or inter-hemispheric oscillations, comprised of various 'mixes' of right- and left-hemisphere functions"; (e) the overall "acuity" of right- versus left-hemispheric functions may vary across cultures (as well as subcultures, social classes, and individuals); (f) such variations reflect learning that different cognitive strategies have different survival potentials; (g) habituation to one type of cognitive strategy does not preclude adoption of another type of cognitive strategy; (h) the two basic modes, or strategies, are evaluated differently across cultures: " ... the best example is the tendency of Westerners to regard only what appear to be manifestations of left-hemisphere functions as 'real' intelligence".

Paredes and Hepburn oversimplify the results and implications of the research on hemispheric functions of the brain, but their suggestions are in line with other attempts to relate this research to problems in the social sciences ranging from cross-cultural philosophical differences to the sociological nature of scientific inquiry. Consider, then, the extent to which Needham's findings manifest a relational way of thinking rooted in a relatively uncarpentered, agriculturally dominated landscape. To what extent is such a situation suggested in the characterization of Chinese algebra as beautifully "symmetric", "rhetorical", and "positional", not to mention the dominance of algebraic over geometrical orientations in Chinese mathematics? Are the high Phenomenal Regression Index of the Chinese, eidetic artistry, and the use of parallel perspective further evidence for a Paredes–Hepburn type of hypothesis? Are these phenomena simply pre-modern patterns of thought which may or may not be significantly related to hemispheric functions in the brain; and if we are dealing with a general pre-modern phenomenon, is there reason to suppose that environmental and cultural conditions in China were peculiar in the extent to which they fostered the development of an appositional mode of thought?

Needham refers again and again to phenomena which seem to make sense in terms of the Paredes–Hepburn hypothesis. Thus, he refers to (a) the interest among Taoists in techniques which could not be transmitted by words; and (b) the traditional Chinese shipwrights who depended on the skill and eye of the older and more experienced shipwright rather than on templates or blueprints. These examples have their parallels in medieval Europe and other pre-modern cultures; but the Paredes–Hepburn hypothesis is not thereby eliminated, since the evidence Needham cites suggests a relative and not an absolute difference – a small enough difference perhaps (recalling Campbell) to give rise to Needham's Problem of why modern science emerged in Western Europe and not in China. Of course, the Chinese are, like most people, predominantly right-handed. But split-brain advocates might note that the significance of the left (suggesting right-brain dominance) in Chinese symbolism is a relatively unusual cultural phenomenon. This speculation is intriguing but fraught with pitfalls. It seems more reasonable to assume that different environments and cultures elicit and educate different cognitive abilities. The differences between China and Western Europe are much more likely to be discovered in environmental and sociocultural differences than in any inherited tendencies toward left or right brain dominance.

The heuristic value of the speculations I have reviewed depends on the extent to which Needham and others have "accurately" portrayed Chinese culture. The speculatons also specify areas of concern for scholars trying to understand human beings in a cross-cultural perspective. In general, there does seem to be some agreement among students of Chinese culture on Needham's characterization of the dominant mode in Chinese thought as "organismic". There is, however, a question about whether this mode is Whiteheadian, Needham's label for a phenomenon that other investigators would recognize as pre-modern, or something altogether different.

SECTION II

THE SOCIOLOGY OF PHYSICS AND MYSTICISM

CHAPTER 3

MYSTICISM AS A SOCIAL FACT

INTRODUCTION

Parallelists view mysticism as a mode of thought that allows us to experience reality "directly". They accept the mystical experience in and for itself. For example, Capra contends that Eastern mystics have always realized "that reality transcends ordinary language". He quotes Ananda Coomaraswamy in support of the idea that the mythical language used by mystics "embodies the nearest approach to absolute truth that can be stated in words".[1]

Parallelists stress the profundity, purity, and perfection of mystical knowledge. In effect, they immunize mysticism against the methods and theories of critical inquiry. Thus, Capra argues that "mysticism is, above all, an experience that cannot be learned from books". More generally, parallelists (a) underscore the mystery of the *similarities* in reports about mystical experiences by people who have had no contact with one another, and (b) overlook the *variety* in mystical experiences. Capra views Eastern mysticism as "a vast number of subtly interwoven spiritual disciplines and philosophical systems"; but he asserts that the basic features of the world views in Hinduism, Buddhism, and Taoism are the same. Furthermore, the "universal" experiences of mystics are, according to Capra, basically religious; mysticism and religion are inseparable.[2]

The parallelist conception of mysticism is an extension of the mystification, glorification, and worship of mystery by such mystics-scholars as E. Underhill, D. T. Suzuki, C. Humphreys, and R. Jones. Arguments in defense of, and descriptions of, mystical experiences are offered with varying degrees of missionary zeal. Suzuki is exemplary in this respect; and this is of special significance because of his pre-eminence as an interpreter of Buddhism (especially Zen) in lay and scholarly circles. Suzuki stresses the notion that the uniqueness of Zen "lies in its irrationality or its passing beyond our logical comprehension".[3] His views on mysticism are widely quoted throughout Capra's book.

There are many variations in the mystification of mysticism. Sometimes mysticism is associated with a perspective on the future of religion and spirituality. Thus, S. Radhakrishnan conceives mysticism as "the admission of mystery in the universe". He refers favorably to Bergson's view of mysticism as "open religion", and predicts that mysticism, standing for "the fellowship of humanity" and being the "religion of the spirit" that avoids the dogmatic extremes of affirmation and denial, "is likely to be the religion of the future".[4]

In some cases, mystification is used to buttress an argument against conventional objectivity and rationality in science. Roszak, for example, argues that there is an immense "psychic distance" separating the scientific view of the uniformity of nature from the Oneness of the Old Gnosis. He affirms the sacramental vision of the Old Gnosis as an alternative to "objective science". The nearest objective science comes to the Old Gnosis is ecology, and Roszak (oblivious to Yates' interpretation, of which more later) finds its "founding fathers" in the Hermetic philosophers and Romantic poets. That is not sufficient; he goes on to trace the origins of ecology back to "the Tao". In his eloquent plea for "a great restoration" (or "apocatastasis"), Roszak is constantly drawing us "back" to mysteries: the restoration cannot come about through argument or pleading, but only of "its own mysterious accord"; the "debunking spirit of science" is taken to task because it represents "a nihilistic campaign against the legitimate mysteries".[5]

The parallelists take a similar stand on the mysterious. LeShan, ignoring anthropological hypotheses about situational determinants of parallel inventions, discoveries, or experiences (as well as the pitfalls of comparative linguistic analysis), describes the similarity of reports on mystical experiences – "including reports of those who clearly had no contact with each other" – as "overwhelming".[6]

One of the most telling statements in Capra's defense of parallelism is the following:[7]

> ... the principal theories and models of modern physics lead to a view of the world which is internally consistent and in perfect harmony with the views of Eastern mysticism.

The phrases "internally consistent" and "perfect harmony" suggest a psychological search to satisfy a personal need (one which may very well be widespread and explicable as a sociocultural phenomenon), for (a) there is no physical theory which *in itself* is, or can long remain, either internally consistent or perfectly harmonious (in the sense that this implies finality in terms of theoretical development); and (b) if there were such theories they certainly could not be expected to remain unchanged and one would then anticipate – since Eastern mysticism would remain *relatively* stable (changing perhaps as we continually reconstruct and reconstrue history based on new levels of awareness, consciousness, and comprehension) – that the parallels would sooner or later disappear!

Capra also writes about the "profound insights" of the outstanding Hindu teachers; and he argues that Hinduism " . . . almost completely determined society and culture in India", a one-sided, overly-deterministic view he quickly modifies to allow for social influences on Hinduism. But in general Capra ignores the social roots of ideas and activities.[8]

Capra affirms a basically asocial view of "the spiritual battle of man", represented by the battle of Arjuna in the Bhagavad Gita – "the warrior in search of enlightenment". And if one tries to find a social message here it turns out to be one which I doubt Capra would like but one which nonetheless is suggested by his images and affirmations: the social transform of the warrior in search of enlightenment is a variety of fascism (this connection will appear more plausible after I have reviewed Forman's study of physics and culture in Weimar Germany).

The "unlearnable" nature of mysticism (presumably in the sense of socialization and cultural heritage) is stressed by Capra and other parallelists, and contributes to the mystification of mysticism. At the same time, physics is made rather easily (but falsely) accessible.[9]

The problem of mystification is by no means restricted to mysticism. *Mathematics* has been conceived as a mystical quest, and mathematicians have been *accused* of being mystics. In the Renaissance, "Genuine mathematicians" debated those with "Pythagorean intentions" (e.g., the Kepler–Fludd debate); today certain branches of mathematics, such as catastrophe theory, are branded idealist and Pythagorean. Mathematics and mysticism are linked repeatedly (one might say endemically) in the history of thought. Their relationship is relevant to the present discussion but cannot be pursued further here. The reader is referred to the seminal analysis on numbers and religion by Oswald Spengler discussed in Part II.[10]

The emphasis on "pure" experience in mysticism is difficult to account for unless we assume that it is a strategy in intellectual conflict, and/or a reaction to some sort of threat. Both mystification and purity can be regarded as responses to real or imagined threats. I will say more on this later. People in all cultures, in any case, apparently *experience* things that are "awe-inspiring" or "extraordinary". The *interpretation* of such experiences is not, however, a universal. Mystery can be and is approached from a naturalistic perspective in some cases, and with uncertainty and wonder (as *static* intellectual orientations) in others.[11]

DEMYSTIFYING MYSTICISM

The sociological demystification of mysticism is part of the broader program of the sociology of religion. While the sociology of religion is a recognized specialty, the sociology of mysticism is more of an hypothesis than a field of inquiry. It is not difficult to locate ideas on the social aspects of mysticism. Weber, perhaps, had more to say about the sociology of mysticism than anyone before or since his time. Among the ideas about mysticism and society proposed by sociologists who came after Weber are (a) Gurvitch's suggestion that mystical knowledge (which plays a role in all *types* of knowledge) triumphs over all other *forms* only in archaic societies, (b) Stark's less specific comment on the neglect of mystical experience during the last few hundred years, and (c) Scheler's association of "light" and "dark" forms of mysticism with, respectively, patriarchal and matriarchal social orders.[12]

Anthropologists have made many important contributions to the study of mysticism. The overall impact of this research has been to add substance to E. B. Tylor's blunt and pithy "theory":[13]

> Bread and meat would have robbed the ecstatic of many an angel visit: the opening of the refectory door must many a time have closed the gates of heaven to his gaze.

The emergence of Christianity, Judaism, and Islam has been correlated with uprooting changes in social and economic conditions. In Africa, the "theory" of spirit possession and the emergence of individuals as religious specialists has been found to be associated with experiencing "long or repeated illness, recurrent dreams, hysterical fits and trances, and other violent, abnormal behavior". Bridges defends a parallel *societal* thesis in his analysis of mysticism in America. He argues that war and social turmoil demonstrate the transience of the material world and turn people toward "inner-seeking". This "catastrophic" view does not invalidate but may obscure the "gradualist" view defended by Geertz. Geertz argues that there is a constant background of incomprehensible aspects of nature, and problems of suffering, evil, and meaning "which renders man chronically uneasy and toward which a more equable flow of diagnostic symbols is consequently directed".[14] Durkheim's sociology of religion offers some interesting insights on these issues.

Durkheim's basic premise is that "religious beliefs rest upon a specific experience whose demonstrative value is, in one sense, not one bit inferior to that of scientific experiments, though different from them". Specifically,

Durkheim hypothesized that religion is the image of society, reflecting "all its aspects, even the most vulgar and repulsive": [15]

> All is to be found there, and if in the majority of cases we see the good victorious over evil, life over death, the powers of light over the powers of darkness, it is because reality is not otherwise. If the relation between these two contrary forces were reversed, life would be impossible; but, as a matter of fact, it maintains itself and even tends to develop.

Collective life, Durkheim argued, affects psychic activity. The individual, psychically transformed through the over-excitation of "vital energies", transforms his/her environment. The priority of the collectivity is psychically experienced as a "sacred" world distinguished from and above the "profane" world. This suggests that perhaps at the other extreme from the collective experience, *wandering* and being alone, changes can be expected in psychic states; and – since the collectivity is "physically" absent – the experience is the mystical encounter with the "unspeakable" Oneness. *Is this Oneness related to the individual as religion is to society*?

Consider, to pursue this further, Durkheim's comments on the individualization of collective ideals. [16] Each individual understands collective ieeals

> after his own fashion and marks them with his own stamp; he suppresses certain elements and adds others. Thus the personal ideal disengages itself from the social ideal in proportion as the individual personality develops itself and becomes an autonomous source of action.

Does this help to account for the experience of mysticism? That is, can mysticism be the result of the process of the individualization of collective ideals, facilitated by social conditions which lead to the emergence of wandering individuals, or niches for hermits? This idea is not inconsistent with the view of mysticism expressed by Evelyn Underhill when she writes that "The 'I' is . . . the 'point of departure' ".[17]

These preliminary inquires suggest, in brief, that what we need to carry out a program of demystification is a sociological conception of "Godhead" and related ideas that parallels Durkheim's conception of religion. At the same time, we must heed Geertz's caution against accepting a "vulgar" positivistic view of religion as a *description* of the social order. But we must also be cautious about Geertz's equally vulgar alternative assertion that religion *shapes* the social order.[18]

TOWARD A SOCIOLOGY OF EASTERN MYSTICISM

Following Scholem, who relies on Troeltsch, my working definition of mysticism is "the striving for immediacy, inwardness and presence of religious experience . . . an awareness of the living experience of the Divine, vouchsafed to individuals living in institutionalized and traditional forms of religion".[19] The mystic "seeks an illumination, a knowledge and, moreover, an experience that to him has all the signs of immediacy, of overwhelming certitude and evidence". Sometimes, of course, the mystic seeks to escape established religion; but he/she carries the religious imperative into the mystical quest.

My assumptions are that (a) by combining Durkheim's emphasis on how the social order shapes religion, and Geertz's emphasis on how religion shapes the social order it is possible to construct a sociological theory of religion, and (b) such a theory can be applied to understanding the "mysteries" and to demystifying the mysticism in Hinduism, Buddhism, and Taoism, and to understanding the *need* to mystify. I focus here on India, a choice dictated by (a) the common cultural origins of Hinduism and Buddhism, and (b) the attention given by parallelists to the relations between Buddhism and modern physics. The following discussions of Hinduism, Buddhism, and mysticism in general are not intended to be systematic, explanatory accounts. Rather, my aim is to explore some of the literature on the social roots of mysticism in order to establish a rationale for and identify ideas relevant to the development of a sociological theory of mysticism.

HINDUISM

The complexities, the "wholism", and the "modern" sense of large numbers characteristic of Hindu cosmology have been pointed to by parallelists as evidence of anticipations of modern physics. In fact, these characteristics can be better understood as historically and culturally conditioned, that is, as expressions of the transactions between particular people at particular times with particular environments. This understanding does not remove phenomenal equivalencies – the large numbers of Hindu and modern cosmology remain. But the notion that the phenomenal equivalencies reflect conceptual equivalencies is rejected.

The first and most general point is that the enormity and complexity of Hindu cosmology reflect (a) the enormity and complexity of ancient Indian society, and (b) the tendency in ancient India to juxtapose new and old cultural elements in a developing hierarchic social structure. The systematization of Indian cosmology developed "by aggregation and encapsulation". Gombrich writes:[20]

> ... for instance different cosmologies are generally accommodated by making them occur successively, rather than by, say, interpreting one story as an allegorical alternative to another. It is this, I think, which largely accounts for the notorious fact that the dimensions of both space and time in the classical Indian cosmologies are so unconscionably large; two systems are reconciled by putting the one inside the other, and making it a cosmographical or temporal part of a much larger whole . . . the enormity and complexity are like those of the enormous and complex social system, and exemplify the Indian tendency not to supersede new cultural elements but to juxtapose them with the old in a hierarchic ranking.

The dimensions of space and time in Indian cosmology thus reflect a tendency to relate new and old views of nature and the world additively, rather than to conceive them as allegorical alternatives. Dumont also notes this tendency to rank rather than to exclude, and describes it in terms of hierarchy and complementarity.[21] This is not unlike what Ficino, Pico della Mirandola, and Cornelius Agrippa did during the Renaissance: the culmination of this tendency in Bruno's infinities has led many moderns to erroneously attribute a modern world view to this "magician". I will discuss this in Chapter V.

Dumont's absorbing essay on the caste system, *Homo Hierarchicus*, is consistent with the idea that Hindu cosmology reflects the structure of Indian society. The translation of the structural characteristics of the caste system – directly and indirectly – into the language of religion and mysticism helped set the stage for parallelist interpretations of the relationships between modern physics and Eastern mysticism. The notions of continuity, wholism, complexity, and levels which parallelists find in Hindu cosmology reflect aspects of Indian social structure. Consider, for example, Dumont's view that the set of castes in a definite territory "may be spoken of as a geographically circumscribed system of castes". His discussion of the "jajmani system", the intercaste network of interdependent specializations, and the prestations and counter-prestations by which castes as a whole are tied together in the village stimulates an image of interpenetration and wholism. The hereditary groups which make up the castes are distinguished from one another and connected together (a) by gradations of status or hierarchy, (b) by rules of separation, and (c) by a division of labor. All three of these characteristics of caste "are given together". They rest on one fundamental conception.[22] Given my earlier arguments, caution must be exercised in relating mystical imagery and imagery in descriptions of social structure. Still, Dumont's work suggests the working hypothesis that the rhetoric of wholism and interpenetration that parallelists interpret in terms of bootstrap physics is grounded in the relative lack of institutional differentiation and autonomy in traditional societies and, in the case of India, the persistence of kinship and religious bonds under conditions of increasing *aggregative complexity* in the social structure. Aggregative complexity is the result of "piling" elements one upon another in a random or arbitrary manner (as opposed to complexity based on, for example, a rational or logical ordering, or a progressive development such as evolution). The aggregative nature of ancient Indian social structure is noted in various descriptions of Hinduism and the caste system.[23]

Charles Eliot referred to Hinduism as "a jungle, not a building". He suggested that the Brahmanic "bond of union" was a major factor in creating the "great national paganism" of Indian society. This orientation, and perhaps the other aspects of the Indian thought system referred to earlier, may reflect an adaptive response to waves of invasions. The "bootstrap" and "interpenetration" ideas parallelists make so much of in relating Hindu and modern cosmologies may be a consequence of this adaptive mode. The situation is illustrated by the case of Indra – who is progressively declared to be the same as Agni, and the same as Varuna. Probably every Vedic deity is at some time identified with another deity. Eliot's ethnocentric views did not interfere with his ability to suggest an ecological theory of Hinduism:[24]

> Hinduism has often and justly been compared to a jungle. As in the jungle every particle of soil seems to put forth its spirit in vegetable life and plants grow on plants, creepers and parasites on their more stalwart brethren, so in India art, commerce, warfare and crime, every human interest and aspiration seek for a manifestation in religion . . .

Weber has numerous insightful hypotheses on the sociocultural roots of Hinduism. In line with the current discussion, he suggests that the "profusion of numerical schemata" (a factor which is related to the "large numbers" aspect of Hindu thought) is a consequence of the oral tradition among the Brahmin priests. Versification was also part of this tradition, and must be considered a factor in creating the possibility for wide-ranging modernist interpretations by today's parallelists. As Bendix notes:

> Easy comprehension and ready reproducibility were imperative. Hence the oldest philosophical writings and the sutra literature (dealing with the ritual practices of everyday life) are written in the form of epigrams that can be learned by heart and commented upon by the teacher. For the same reason much of the nonphilosophical literature is written in verse form. Statements of an idea or prescriptions for a ritual observance are repeated endlessly, each time with the modification of only one word or sentence, in keeping with the sequence of thought.

The crucial point is that these characteristics, which very likely originated in efforts "to facilitate memorization", led

> to an endless accumulation of embellishing adjectives, comparisons, and symbols, as well as to endeavors to *simulate* [my emphasis] an impression of the great and the godly through the use of gigantic numbers and luxuriant fantasies.

The oral tradition is a crucial factor in understanding the development of Hinduism. Disputations, debates, and forensic contests were important in Hindu culture. These activities depended on easy comprehension and reproducibility. The epigrammatic form is well-suited for this; it makes it easy to learn things by heart, and it provides a basis for the teacher's commentary. It was not until the second century AD that the learned Brahmin men became a class of scribes. The demands of oral traditions and ritual may account, along with other related factors, for the "profusion of numerical schemata and the pedantically detailed outline of the material". The "penchant for rhetoric and embellishment" stemmed from and reinforced a "studied indifference (non-empirical orientation) toward the world".[25]

The cultural roots of Hindu large-number systems is suggested by changes in those systems. Jaki argues, for example, that "the time-scale of cosmic and historic cycles was the outgrowth of a shattering disappointment with the turn of historic events". In some of the older Hindu writings (e.g., the great epic poem, "The Mahabharata"), the four yugas have a total span of only 12 000 years; in later writings, the span reaches in excess of four million years. Caught up in the frustrations of daily living, those who sought escape through detachment, through working to develop a sense of the illusory nature of everyday life, fixed their minds "on the endless recurrence of the same hopes, same frustrations, same triumphs, and same defeats. As the imagination of each generation of Hindu sages kept adding its contribution, the final picture became gigantically mesmerizing".[26]

BUDDHISM

In arguing for parallels between Buddhism and physics certain developments in the history of Buddhism must be taken into account. It is important to remember that Buddhism originally proscribed "solitary asceticism", and "efforts to attain sublime heights of knowledge":[27]

> It had no occult, no esoteric system of doctrine which it withheld from ordinary men.

Most importantly, it did not at first concern itself "with any form of philosophical or metaphysical teaching". The teachings that are seized on by parallelists are the product of religious speculations carried on by upper-class individuals of leisure. This is true of the two great Mahayana schools, the Madhyamika and the Vijanavadin (Idealistic), both of which appeal to the mind rather than the heart.[28]

Buddhism emerged during a period (the seventh and sixth centuries BC) of deterioration and dissolution of old forms of social organization. During this period, spiritual malaise reinforced earlier thinking about the nature and meaning of self and world. Traditional ways – during this period of "suffering" – were questioned more sharply than ever; and the world of everyday life was radically devalued. The Buddhist emphasis on personal spiritual independence was clearly adaptive. Pardue uses the term "therapeutic". The situation is exemplified in the towns of the Ganges Valley; this is where Kapilavastu (80 miles north of Benares) is located, the town in which Buddha was born about 560 BC. Every year the river flooded, destroying crops that had demanded struggles with a harsh earth. The yearly monsoons brought famine, drought, dysentery, cholera; weakness made people easier prey than normal for predators like the tiger and the snake. In the face of all of this, the Hindu priests (Brahmana) chanted Vedic hymns and offered fire sacrifices to Brahma. Monier-Williams and Margaret Stevenson both comment on climate as an important factor in generating pessimistic viewpoints. Stevenson argues that people were prompted to seek the life of the professed ascetic by the combination of creed, climate, pessimism, and injustice.[29]

The Buddha and Mahavira (the founder of Jainism) were both from the upper-class; and furthermore, they were from indigenous non-Aryan tribes. Their teachings were warmly received by "upwardly mobile urban commercial and political groups, who held that both religious and social standing should be based on personal talent and achievement rather than hereditary right". The Buddha was a prince. But his actual circumstances were humble. I can only speculate, with Pardue, "on the irrelevance of the warrior model and the archaic cult to the prince of a small enfiefed clan in an age of brutalizing warfare".[30]

The social circumstances which stimulated the development of Buddhism and made it eventually attract contemporary parallelists are suggested by Pardue:[31]

> In a sense, the subsequent history of Buddhism centers on the possibilities for exploring the implications of the teaching developed in articulation with the immense spectrum of discrete human needs, institutions, and cultures with which it was brought into contact.

It is not improbable, furthermore, that the lack of a system of authority that could be brought to bear on the Buddhist monks (such as the one found in the Western Church) encouraged a freedom of spiritual inquiry and contributed to the intellectual virtuosity of Buddhist thought. Not long after his first sermon, sixty monks from the upper classes were "enrolled" with the Buddha. The sixth to be admitted to the Sangha was a high-born youth named Yasa. His father, a rich merchant, was the first lay-disciple, his mother and wife the first lay sisters. There were four high-born friends of Yasa who followed, and then some fifty others joined.[32]

Westerners have adapted Buddhism to a wide range of interests.[33] Spiro's analysis and critique of Theravada Buddhism in Burma shows that adaptation also characterizes the development of Buddhism in the East. Theravada (Hinayana) Buddhism presents the anthropologist with a stunning problem, according to Spiro: why is Theravada Buddhism atheistic, where other religions require supernatural assistance for salvation; why does it reject the soul common to other religions; why does it teach "life is suffering" where "most people agree that many experiences afford pleasure and even true happiness"; why does it insist that salvation depends on extinguishing desire, where "most people" believe that satisfying desire is "one of the marks of salvation"; and how is it that it "demands emotional detachment and the renunciation of the world, where most people seek attachment to friends and loved ones, power, and position, self and possessions"?[34] This problem, with the "remarkable opportunities" it affords to discover sociocultural conditions that lead to the "acquisition of beliefs and attitudes which constitute dramatic exceptions to our empirical knowledge and our theoretical expectations concerning human nature and religion" is a pseudo-problem. Anthropological studies of living Buddh*ists* show that they are not all that different from other people, though we would conclude differently based on textual Buddh*ism* and treatises *on* Buddh*ism*. Spiro distinguishes two normative religious systems: the *ordinary norm* is for the religious majority; the *extraordinary norm* is for the smaller group primarily concerned with salvation. This distinction applies to ancient Indian Buddhism and Hinduism, to Theravada Buddhism, and to Mahayana Buddhism. Similarly, Suzuki distinguishes *practical* from *speculative* Mahayana.[35] The distinction is applicable to all religious systems. The importance of the religious virtuoso in Buddhism (as well as in Hinduism and in other religious systems) as the source of the extraordinary norm must be recognized. This suggests a way of widening the earlier perspective I sketched, following Weber, on the sociocultural roots of pessimism and its elaboration in the widely noted unbridled speculations, imaginings, and inventions of the Buddhist virtuosos.[36]

Suzuki argues that Hinayana and Mahayana are not as different as some people imagine. The differences are due to (a) the general development of religious consciousness and the broadening of intellectual horizons, and (b) "the conservative efforts to literally preserve the monastic rules and traditions". He notes further:[37]

In ancient India everybody of the higher castes seems to have indulged in intellectual and speculative exercises. Being rich in natural resources and thus the struggle for existence being reduced to a minimum, the Brahmins and Ksatriyas gathered themselves under most luxuriously growing trees, or retired to the mountain-grottoes undisturbed by the hurly-burly of the world, and there they devoted all their leisure hours to metaphysical speculation and discussion.

We can readily believe, Monier-Williams writes, that in a man like Buddha, "introspection and abstinènce, protracted for many years, induced a condition of mind favourable to ecstatic visions, which were easily mistaken for flashes of inner enlightenment":[38]

We know, indeed, that eleven centuries later another great thinker arose among the Semitic races in Western Asia, who went through the same kind of mental struggle, and that Muhammed, like Gautama, having by his long fasts and austerities brought himself into a highly wrought condition of the nervous system, became a fanatical believer in the reality of his own delusions and in his own divine commission as a teacher.

Buddhism may be considered a virtuoso religion in conception, with the monk as the Buddhist par excellence. This is virtually acknowledged by the Buddha who says that while it is *possible* for a householder to live as a true Buddhist, it is extremely difficult. Renunciation – and thus the monkish life – is, he says, most conducive to the Buddhist life.[39]

Consider next the emergence of the *bodhisattva* ideal in Mahayana. Originally, this referred to previous incarnations of the historical Buddha. It became universalized to encompass persons worthy of nirvana but committed to sacrificing this in order to "give love and compassion to all sentient beings as a condition of . . . salvation". This opened the teachings of Buddha to "new symbolic forms, beliefs, and practices", as Pardue notes:

It facilitated popular diffusion and provided the basis for theistic and philosophical developments reflected in the principle Mahayana texts and schools. Equally important was the overt affirmation of the divinity of the Buddha. He is not only the historical

teacher; he is an omnipresent deity, an eternal spiritual being and active force. This allowed for the development of a theory of continued revelation with innovative doctrinal and institutional consequences.

Other factors important in the intellectual evolution of Mahayana were (a) the powerful intellect of Nagarjuna (ca. 200 AD), founder of the Madhyamika school, and (b) the idea of "upayakaysalya" (Buddha's "skill-in-means" – "the ability to adjust teachings and institutions to the needs of all sorts and conditions of men"; this operated in missionary diffusion, and "allowed for almost unlimited adaptability to given conditions".[40] By the 9th century AD, the Mahayana philosophical schools "became increasingly preoccupied with abstruse theoretical issues centered around problems of logic, epistemology, and perception. In time, theistic Mahayana and Tantric Buddhism became hardly distinguishable from the increasingly luxuriant garden of Hinduism".[41]

The tendency to see Buddhism as a "psychology" is widespread among modern, especially Western, interpreters of and adherents of Buddhism. In arguing for the "real concerns", and the "real content and work" of Buddhism, Watts expresses dismay at the propagation of Buddhism in the West as an organized religion. But his argument that the "real concerns" of Buddhism "are closer to psychotherapy, or even to something such as opthalmology, than to the choice between differing systems of belief which we recognize in the West as adopting a religion" masks his personal adaptive orientation to Buddhism. Spiro admits the possibility of investing Buddhism with psychological meaning, but he is cautious about doing so. The third of the four noble truths, for example, is that since suffering is caused by the frustration of desire, the cessation of suffering can be achieved by extinguishing desire. This has been interpreted by some people as a "naturalistic psychology". Spiro argues that this is a "half-truth" at best. The statement that existinguishing desire can eliminate suffering is – in the Buddhist context – an *ontological* rather than a psychological assertion. The basic message is an eschatological one. Desire causes rebirth, that is why it is the root of suffering; extinguishing desire eliminates suffering because it releases human beings from the Wheel of Rebirth.[42]

The basic distinction that Suzuki discusses in great detail between *prajna* and *vynana* is the source of additional misconceptions about Buddhism and psychology. Distinctions such as that between prajna and vynana are documented in studies of cognition and culture and variously referred to as non-linear and linear modes of thought, appositional and propositional modes, coordinative and subordinative modes, and so on. There may be conceptual difficulties associated with the development and application of such categories, but there is certainly no "mystery" about them. Yet Suzuki has suggested otherwise. He has argued that the only thing we can say about Zen in general is that its irrationality makes it unique and beyond our rationalistic or logical ken. Hu Shih objects to this viewpoint:[43]

> My greatest disappointment has been that, according to Suzuki and his disciples, Zen is illogical, irrational, and, therefore, beyond our intellectual understanding.

By contrast, Hu Shih argues as follows:

> The Ch'an (Zen) movement is an integral part of the history of Chinese Buddhism, and the history of Chinese Buddhism is an integral part of the general history of Chinese thought. Ch'an can be properly understood only in its historical setting just as any other Chinese philosophical school must be studied and understood in its historical setting.

Hu Shih refers to the period from 700–1100 AD as the "age of Ch'an". Documents indicate that during the first 150 years of this period the great masters (Shen-hui, Ma-tsu, Hsuan-chien, I-hsuan, etc.) "taught and spoke in plain and unmistakable language and did not resort to enigmatic words, gestures, or acts. Some of the famous enigmatic answers attributed to Ma-tsu and his immediate disciples were undoubtedly very late inventions". New situations arose which "led to the development of a pedagogical method of conveying a truth through a great variety of strange and sometimes seemingly crazy gestures, words, or acts". One was that the Ch'an schools became "respectable and even fashionable in intellectual and political circles", and numerous monks and lay dilettantes emerged who spoke the language of the masters with little understanding or conviction. The term "ch'an of the mouth-corners" (K'ou-t'ou ch'an) refers to the deterioration of the great masters' teachings.

I-hsuan probably introduced the "crazy" techniques of later Ch'an Buddhism. He was famous for beating his questioners with sticks, and shouting loudly at them. But, Hu Shih suggests, underlying the superficial irrationality was a pedagogy. Basically, this pedagogy represented an attempt to educate by "the hard way", by allowing (prompting) individuals to learn through their own efforts and "ever-widening" life-experiences. There are, Hu Shih argues, three stages in this pedagogy: (a) pu shuo p'o: never tell too plainly; (b) answer questions eccentrically (by beating with a cane, remaining silent, presenting paradoxes, and using non-sequitors such as "go wash the dishes"); (c) hsing-chao: "traveling on foot". In stage three, the novice who has failed to understand the master's answers is told to leave, goes off on foot with a stick, a bowl, and a pair of sandals, and begs for food and lodging. Under these conditions a "trifling incident" is supposed to bring Sudden Enlightenment.

Hu Shih need not be correct in all particulars; his method, however, is compatible with a sociology of knowledge point of view. Suzuki's replies are unacceptable and indicative of why Eastern mysticism becomes mystified.

The problem of advocacy and adaptation in Buddhism can be seen in considering the issues raised by Hu Shih somewhat more broadly. Benz dates to "the day and hour" the penetration of the American scene by Asian religions: the speech by Anagarika Dharmapada on Theravada Buddhism to the World's Parliament of Religions at the Chicago World's Fair in 1893. Soyen Shaku also spoke there, on Zen Buddhism.[44] Suzuki had "deep roots in the Buddhist monastic tradition"; yet he "did more than any other to shake Zen loose from its Buddhist roots and thereby to make Zen acceptable to Westerners – going so far as to procliam that "Zen of itself proclaims to be the essence of Buddhism; in reality it is the essence of all religions and philosophy". Indeed, Suzuki was a disciple of Soyen Shaku, a leader of the Rinzai Zen sect. He was one of three disciples specifically chosen by Shaku to *promote* Zen in America. Another widely read interpreter of Buddhism, Christmas Humphreys, has been described as Britain's "most articulate and voluble Buddhist".[45]

Two important aspects of Buddhism, and their social correlates, can be identified based on the preceding discussion along with some further considerations. One is *wandering*. The connection of wandering to the mystical experience can be gathered from a review of the descriptions of mystics from Buddha to Giordano Bruno, to the nomadism of Castañeda's, Don Juan. The other aspect, widely noted, especially in Weber's work, is *virtuosity*. Where wandering is the first act of the person who turns to the mystical life, especially because he or she must renounce the world, then – at least for the well-to-do – monastic institutionalization and the development of a virtuoso religion are likely to be social consequences. This then sets the foundation for elaborating religious notions and theology without end. And as Yates notes in considering what Giordano Bruno – wanderer and virtuoso mystic – did in developing Ficino's views as set out in *De vita coelitus comparanda*, " . . . once you start a religion, there is no knowing what it may become".[46]

Pravrajya, samnyasa ("the complete casting off"), and agarasma anagarujam ("passing from home into homelessness") are associated with the rejection of the Vedic tradition and of the world, and of the ritualistic "going forth". By the sixth century BC, it appears that followers of this way of life had grown into a community of wandering religious ascetics, paribiajaka:[47]

> They live by begging, have no settled dwelling (except during the rains, when the observance of the rain retreat is a common custom among them), move about from place to place, and are either ascetics practicing austerities or are, in the words of Rudyard Kipling, "dreamers and babblers of strange gospels".

These were the propagators of Buddhism: "strictly contemplative, mendicant monks, who rejected the world and, having no homes, migrated".[48]

In society, the renouncing mystic would develop a conventional view of God, as a projection of "Society" transformed into an abstraction free of the trials, tribulations, demands, and sufferings of everyday group life. *Outside* society, in the "going forth", there is no "social soil" for such people to *root* their "feelings" in. The root-less feeling is interpreted as a mystical experience; and it may, as Weber argues, follow from seeking salvation through a religious conviction that is hostile to society. The aim is to be a "vessel" (Gerfass) of God, to find rest in God.[49]

Weber also suggests a "politics of mysticism" when he notes the "affinity" between concepts of God and political organization. He argues, for example, that "Involvement with the techniques of civilization militates against religious innovation"; lack of such involvement encourages religious innovation. The mystic, therefore, works out innovation in a virtuoso manner. The "alternative politics" of the mystic is reflected in his/her religious viewpoints.[50]

I do not want to obscure or ignore the fact that not all mystics renounce the world, nor to correlate renunciation uncritically with the world of pure physics as a basis for understanding affinities underlying parallelism.[51] However, parallelism does seem to draw together the world of the pure physicist and the world of the virtuoso mystic. The otherworldliness of Brahmins, Buddhists, and Taoists is not unrelated to the fact that in the East, for reasons too numerous to outline here, traditional social organization persisted for a longer time than in the West. Status groups such as those of thinkers and prophets in the West did not develop "because of the enduring power of kinship group, the dominance of caste differences (in India), and the centrally organized officialdom that could gain power in the absence of autonomous and politically effective group formations (in China)". This allowed Buddhists, for example, "extra time" in a sense to work out (and complicate in virtuoso fashion) their belief system.[52]

Standing "outside society", the mystic – in need of authority just as much as some non-mystics *in* society – turns to *experience*. That is all there is to turn to. And this experience – given a need to be subject to some form of authority – tends to get translated into God or a god surrogate.

The ancient notions, far from being anticipations, must be understood as elaborations, decorations, symbols games, abstractions of abstractions, metaphors of metaphors, and logical unfoldings of quite ordinary experiences not unlike those that mark the movement of arithmetic from the realm of common experience into the realm of abstract mathematics. A comparable abstruseness characterizes systems of thought that have developed in a "complexifying" manner, and only the imperative of drawing and testing conclusions about common experiences (common "realities") saves certain forms of inquiry (including some forms labeled "science") from becoming, like mysticism (the notion of Buddhist "experimentation" notwithstanding) "unmentionable", "ineffable" experiences.

A NOTE ON THE POLITICS OF MYSTICISM

Very little research has been done on the politics of mysticism. However, Collins' study of the occult and mystical traditions offers evidence that mystics throughout history and in various cultures have struggled among themselves and with other intellectuals for privilege, wealth, and prestige. They have manipulated ideas in order to further their interests (a thesis I discuss more fully in Chapter 5); that is, they have used ideas as resources in struggles for power. Furthermore, Schutz and Horton have alerted us to the ways which have been developed to account for failures and errors in science and religion, and in physics and mysticism. Each of these realms of knowledge possesses a certain degree of "conceptual resiliency" which immunizes it against external criticism. The power and success of their cognitive defense systems, however, depend on social factors which are less secure from attack by rival systems. Nowotny identifies two groups of such factors: (a) those which safeguard the system and demarcate it, and (b) those "general socio-economic changes which affect the broad structure of risk".[53]

Collins' principles of *occultism* can be adopted here as part of the material for fashioning a sociological theory of *mysticism*. To paraphrase Collins: (a) the mystical traditions are chapters in intellectual history and subject to processes of growth and decline; (b) mysticism is not a single, well-ordered realm; it is "a collection of experiences, hypotheses, and puzzles"; and (c) the mystical traditions contain philosophies, consciousness transforming techniques, physiologies, cosmologies, spiritualisms, and parapsychologies. When we consider such things as the arrogance and aristocratic connections of the Ch'an monks in China, and the military power of medieval Japanese monks, it becomes plausible to view the "multiplicities and conflicts of the mundane world and the unity of mystical heights" as phenomena which are more closely related than commonly thought. Mystical and occult traditions, especially when they become transformed into religions, are implicated in struggles for power, privilege, and prestige.

Thus, we have a situation not unlike that in which science was presented as a "pure" search for truth, a presentation made relatively inaccessible by the mysteries and esotericism of the scientific style of thought and language. Only recently has this view been penetrated, and science brought into the all-too-human world. The image persists, however – and the *need* for the image persists. Mysticism is now being drawn into the inner-circle of science, complete with mystery and esotericism, to help restore that purity and impenetrability.

THE GENERAL ARGUMENT FROM THE SOCIOLOGY OF RELIGION

Religion, according to Worsley, is a process of putting into day to day practice a system of beliefs. It is neither an "ideal operation in the soul, heart, or intellect" (this dymystifies the aristocratic foundations of mysticism); nor is it a "ritual dance of puppets determined by the morphology of their social structure". The critique of Durkheim implied here is not to the point except insofar as the Durkheimian model is interpreted to entail ritual acts and puppet motives, and to be incompatible with reciprocal transactions between individuals and social structures. More generally, both Durkheim and Worsley point to religious activities as adaptations. One should not be surprised that Bachofen failed to reconstruct ancient social organization on the basis of ancient beliefs. Literal parallels are unrealistic (a) because of the interaction between thought and *actual* social organization, and (b) because specialization – especially in the extreme, as in monastic Buddhism – leads religion away from literal mirroring. Certainly, however, religions everywhere and always reflect in some sense their surroundings and the possibilities therein for development and change. Supernatural powers always resemble humans (anthropomorphism), animals (theriomorphism, zoomorphism), values (anthroposocial), or human mentality (anthropopsychic). I cannot agree with Worsley that religion "is neither intrinsically conservative nor revolutionary"; at least not so long as we have nothing better than a general definition of religion as a belief system focused on transcendent meaning. To the extent that one or more central beliefs in a religious system are, for whatever reasons, exempt from criticism and change religion is conservative. If there are *no* dogmas, and certain related conditions are met, then we have a system of *open inquiry*, not a religion.[54]

Consider next the consequences of viewing ordinary language as the raw material of all the different language games humans play. MacCormac argues that the language of science and the language of religion are members of the same family of languages; and they can be relatives because (a) they both use common language and (b) they both depend on metaphors to suggest new meanings. He goes on to reach an important conclusion:

> How could one justify the existence of God by presenting a religious experience that was described as the presence of God? Had one not already assumed the existence of God when he described the experience as "the presence of God"? The charge of circularity will stick only if such invocations are interpreted as proofs or arguments for the existence of God rather than as descriptions of experience.

Myth (and mystification) result both from associating religious or other "ineffable experiences" with the "real world", *and* from believing that rational and mathematized theories *actually* describe the "real world". MacCormac sounds a warning similar to that sounded by Bohm. We should avoid myths because they mistakenly treat "highly suggestive and tension-filled" root metaphors as literal descriptions of reality. We do not have and cannot anticipate having complete and absolute knowledge immune to change.[55]

The mystic learns ordinary language prior to becoming a mystic. This idea reinforces some earlier arguments. The mystic is a socialized being; he/she has "internalized Society"; his/her aloneness magnifies the earlier experience of the "group" and *intensifies* the experience of "God" or oneness (or "zero-ness") that is, according to Durkheim, the basis for religion. It is the *intensification* and *focusing* of the experience of the "collective consciousness" that produces the mystical experience.

We arrive finally at the idea that mysticism, like religion, and science, and all other human activities, is an adaptation. In general, the mystic seems to take the universal problems of chance, suffering, and evil more seriously than other people; and so seriously in some cases that considerations of everyday life are eluded and escaped. And if the mystic interprets his/her experience as one which has "apodictic force", so other religious persons operate on the principle that "he who would know must first believe".

And here, insofar as science is conceived as a Search for Meaning, or Truth, and insofar as it is a substitute in this straightforward way for the religious quest (a functional alternative), then we see again why there can be an affinity between experiences which are labeled *science* and those which are labeled *mysticism.*[56]

Sociologist J. Milton Yinger, and Buddhism scholar Lucien Stryk both suggest that the appeal of Eastern mysticism in the West is based in part on the demythification of the individualistic ethos. Yinger's argument is more general; it relates the emergence of a "countercultural epistemology" to an "unmasking" tradition in Western thought:

> Do not trust appearances or presumed objective truth, for there are deeper realities, say Hume, Marx, Neitzsche, Freud, Pareto, and many others . . . If this unmasking laid the groundwork for the sociology and psychology of knowledge in the thought of some persons, it supported the search for truth through mysticism and the occult among others for whom science itself has been uumasked.

Stryk is more specific:

> Western man's strong response to Buddhism may very well be the result of his being forced by circumstances to accept one of its major premises, that human experience is not individualistic.

Tiryakian carries this argument out in terms of a modernization-secularization hypothesis. His referent is "the occult revival", especially witchcraft. But, even if we distinguish mysticism from occultism, it is not unreasonable to consider Tiryakian's secularization hypothesis applicable. If religion has been secularized, and if – as Tiryakian argues – the demonic has been secularized, then it would be reasonable to assume that mysticism has been – or must eventually become – secularized.[57]

My concern here is shared by at least one student of Eastern religion, H. V. Guenther. He has written an introduction to "The Buddhist Way" based on original Tibetan sources, and under the title, *Tibetan Buddhism Without Mystification*. In the preface, he tells of a complaint he heard from the Dalai Lama in 1961:[58]

> He complained that much of that which purported to be information about Buddhism in Tibet was pure fancy and catered for the mystery-monger rather than for the earnest seeker of religious experience or the critical student of philosophy and that its rich symbolism, especially that of the Tantras, continued to be misrepresented grossly.

The value of mystical experience is emphasized by Guenther for two reasons: (a) it is "the keynote" of Mahayana Buddhism, and (b) it " 'involves' man in actually following [this] path". But he tries to show how this presentation can avoid distortion into mystification:

> ... I have studiously refrained from etymologizing (because we do not understand a work or a proposition by its etymology) and from rediscovering neo-Hegelian trends in Buddhism (although this seems to be the standard procedure by many Easterners and Westerners).

The need for a naturalistic interpretation of mystical experience has been noted by the physicist-philosopher L. L. Whyte who reports having experienced the Quaker "inner light" and the mystic's "unity". In supporting a naturalistic interpretation of these experiences, he writes:[59]

> There is probably nothing about which it is impossible or inappropriate to say anything at all. Furthermore, the connotation of "nature's ineffability" is not that we cannot say anything about it but that we cannot say nearly enough.

A PSYCHO-SOCIAL PERSPECTIVE ON MYSTICISM

A psychosocial theory of mystical experience might be developed by linking C. O. Evans' explanation of mystical experience in terms of states of consciousness, and E. Bourguignon's research on the relations between states of consciousness and social structure. Evans argues that attention is polymorphous, and he identifies four types: (a) interrogative (finding out about something and formulating the discovery propositionally); (b) executive (attending to an activity involving the execution of a technique); (c) unordered (this is a state of "reverie"; "intelligence" is in a sense idle); (d) unprojected (this refers to "those elements of consciousness that together make up the background of consciousness when attention is paid to an object"). Attention is usually polarized, according to Evans, into an object of attention and an unprojected consciousness. Unprojected consciousness is not a homogeneous, static mass; as attention switches, elements are detached from and returned to unprojected consciousness. At any given time, elements "out of attention" may change, vanish, or be replaced. The ability of people like mystics to *claim* they have experienced a "homogeneous consciousness" of one sort or another depends on the human ability to de-polarize consciousness. Thus, it is possible to "attend to the world" in a way which erases the distinction between "object of attention" and "background" (or "unprojected consciousness"). The single element of such a state of consciousness could *not* be an object of attention; such a state might be characterized as one in which "the self" had vanished (if "the self" is identified with "unprojected consciousness"), and for which no *first-hand* report could be formulated at the time of the experience.[60]

If we accept Evans' suggestion that there are *types of attention*, and conceive each *type* as a *state of consciousness*, then *altered states of consciousness* might be viewed as *adaptations* to immediate and to broader sociocultural circumstances. This is indeed the position that Erika Bourguignon appears to have taken. She argues that states of consciousness can be ordered along a continuum varying in terms of central nervous system arousal ("egotropic excitation"). The continuum ranges from increased arousal (relative to a "normophrenic state" characteristic of everyday routine activities) through normal relaxation to decreased arousal. There are two subcontinua: (a) the continuum of increased arousal ranges from hallucination and mystical rapture, through hyperarousal and schizophrenia, to sensitivity, creativity, and anxiety; (b) the continuum of reduced arousal ranges from tranquil states to hypoarousal states such as the yoga state of samadhi.

Each state of consciousness, according to Bourguignon, is an *individual's interpretation of the state of arousal of his or her own central nervous system*. Mystical rapture, then, subjectively symbolizes maximal arousal; samadhi symbolizes minimal arousal. Given states of consciousness (considered solely in terms of how they are induced) may be individual-secular in some societies, and institutionalized-sacred in others. Bourguignon thus significantly elaborates Evans' theory, expands it to encompass psycho- and sociocultural variables, and roots the theory in a series of comparative empirical studies. She concludes that (a) behavior during altered states is subject to a significant degree of learning, and (b) spirits represented in possession trance rituals reveal, to a great extent, symbolic renditions of human society (this reinforces the Durkheimian view I discussed earlier). The possibility of achieving altered states of consciousness, and in particular, "homogeneous consciousness", may vary with the degree of social control exerted on or experienced by the individual. Thus, *wandering* can now be seen more clearly as an important way of breaking conventional social constraints on consciousness.[61]

I. M. Lewis argues that under conditions of deprivation or frustration, people who are cut off from other sanctions will often use possession trance to "exert mystical pressures upon their superiors".[62] Possession trances and mystical experiences may make it possible for individuals to change their social situation without necessarily changing the wider sociocultural setting they operate in. Possession trance in particular, then, should be more prevalent in the more rigid societies. Furthermore, greater use of the mechanism should occur among those who suffer the greatest inability to modify their lives under existing circumstances.[63]

To the extent that altered states of consciousness offer opportunities for a larger number of personal options to the individual within the existing social framework, they help maintain that framework and thus act as a *conservative force*. To the extent that they are linked, for example, to "central possession cults", they are tied to the centers of power and thus again act as a *conservative force*. By providing a sanctioned and prestigious form of decision-making, however, they may, in situations of social crisis, provide an avenue for the expression of dissatisfaction with existing social patterns and for the introduction of *innovations*.

Are mystics agents of change, passive reflectors of change, or reactionaries? Many scholars favor an escapist view of mysticism. But the case of the Buddha is often cited to show that individuals who turn to mysticism are not all despairing, fearful, or social failures.[64] A certain asceticism, however, seems to be a necessary condition for eliciting mystical experiences. And it is difficult to escape the fact of individualistic, if not anti-social, behavior in mysticism. Even when the focus of analysis is on the social aspects of mysticism, individualism stands out.[65] This should not, however, automatically be interpreted as supporting an escapist view of mysticism. To begin with, it might help to distinguish, following Needham, between mystical naturalism (non-obscurantist) and spiritual or religious mysticism:[66]

> All that the former characteristically asserts is that there is much in the universe which transcends human reason here and now, but since it prefers the empirical to the rational, it adds that the sum total of incomprehensibility will diminish if men humbly explore the occult properties and relations of things. Religious mysticism (in the usual sense) is very different; it dotes upon the arbitrary residue and seeks to minimize or deny the value of investigations of natural phenomena.

Mystical naturalism has repeatedly been associated with revolutionary social movements, from those in early Christianity to those in the seventeenth century among the Levellers and Diggers. Taoism was associated with revolts against the established order in China for centuries; Ch'an Buddhism prospered in anti-imperial settings; and Sufism has been characterized as politically local-democratic. But Needham points out that mystical naturalism tends to change into religious mysticism once a new socio-political order is established. Insofar as religious mysticism is prone to mystery worship, it has, I suggest, a volatile potential for surrendering individual power to some form of external power (God, political leaders, cult heroes, etc.). Veysey argues on the basis of his study of mystical and anarchist communes in twentieth-century America that patterns of domination and submission in mystical circles are "fixed with . . . a tenacious hold upon individuals" rivaled only by patterns in "crime syndicates and old-style city politics".[67] The issue is by no means settled. The following hypothesis, formulated by Needham, merits attention:[68]

When a certain body of rationalist thought has become irrevocably tied to a rigid and outdated system of society, and has become associated with the social controls and sanctions which it imposes, then mysticism may become revolutionary.

CONCLUSION

In answer to the parallelist tendency to mystify mysticism, I have identified some of the bases and points of departure for a sociological theory of mysticism. In particular, I have argued that concepts of wholism, interpenetration, and large-numbers in ancient mysticism or wisdom are based on aggregative complexity and numerology, and grounded in ancient social structures and values. The threads of inquiry in this area are not tightly woven; but the rationale for viewing mysticism as a social fact and adaptive strategy is firmly established.

CHAPTER 4

PHYSICS AS A SOCIAL FACT

Capra argues that a "deeper understanding" of the mystical traditions he writes about "can only be felt when one decides to become actively involved" in them; mysticism "cannot be learned from books". He implies that (a) physics *can* be learned from books, and (b) a "deeper understanding" of physics can also be learned from books *without* becoming actively involved in it.[1]

Parallelists claim that physics can be communicated to non-scientists without using mathematics or technical language, and without getting them involved in doing physics. Zukav, for example, argues that "physics is not mathematics"; "Stripped of mathematics, physics becomes pure enchantment". He disputes the idea that learning mathematics is a prerequisite for learning subatomic physics, and seeks support for this view from physicist David Finkelstein. Finkelstein, according to Zukav, contends that it is not necessary to learn mathematics in order to learn subatomic physics. He quotes and paraphrases Finkelstein as follows:

> Mathematics, like English, also is a language. It is constructed of symbols. "The best you can get with symbols is a maximal but incomplete description". A mathematical analysis of subatomic phenomena is no better qualitatively than any other symbolic analysis, because *symbols do not follow the same rules as experience*. They follow rules of their own. In short, the problem is not *in* the language, the problem *is* the language.

But these remarks do not answer the question of whether one has to learn mathematics in order to learn subatomic physics. Zukav's aim – and this is true of parallelists generally – is to somehow make physics palatable to people who are intimidated by, ignorant of, or opposed for one reason or another to mathematical language. Indeed, he asserts that most physical ideas are "essentially simple".[2]

Is mathematics an inherent feature of modern physics? The parallelists say "no", since their goal is to persuade people who do not have mathematical backgrounds that they can understand physics. But physicist Gerald Feinberg has a slightly different viewpoint. He argues that "a *substantial comprehension* of modern physics can be obtained without advanced mathematical training". He assumes that his readers have had a "typical secondary school course" in mathematics. He also notes, however, that "sophisticated mathematics" is essential to theoretical discovery and the practical application of physical theories. Physicist Richard Feynman takes the stronger position that the laws of nature *cannot* be communicated to persons who are ignorant of mathematics.[3]

Some sociologists see physics quite differently than parallelists. The general conclusion to be drawn from a review of studies of science as a social fact is that active scientific fields are heterogeneous arenas characterized by struggles to achieve and establish priority, and conflicts about theoretical, methodological, and substantive issues. Scientists have different interests and are active in different social milieux. The interaction of these differences influences the construction and use of ideas and facts in science. This does not necessarily mean that there are not important areas of consensus in science. However, the parallelists' stress on the *consistency* of modern physics ignores areas of conflict, and the potential for theoretical, methodological, and substantive changes. Thus, Capra refers to "a consistent view of the world" emerging in modern physics; he contends that the concepts of modern physics "all seem to lead in the same direction", and that quantum theory and relativity theory both force us to see the world the same way – that is, "very much in the way a Hindu, Buddhist or Taoist sees it". He stresses the "precision" and "sophistication" of modern experiments, and the "rigorous" and "consistent" mathematical formalism of modern physics.[4] The parallelist argument depends on the various aspects of modern physics converging with each other, on modern physics as a whole developing in a certain direction, and on this development retaining its alleged mystical features indefinitely. All of these notions are at odds with the sociological portrait of science in general, and physics in particular.

The rationale for viewing physics as a social fact is firmly grounded in the works of sociologists of science. These range from Merton's early statements on the dynamic interdependence of science and society to more recent formulations about (a) the interaction between the social structure of science and the "larger structure" of society, and (b) the influence of the sociocultural milieu of a science on the values, beliefs, and ideologies in its disciplinary matrix. There are a number of exemplary case studies of physics as a social fact. Frankel's study of corpuscular optics and the wave theory of light in early nineteenth century France shows that scientists' perceptions of intellectual factors such as diffraction were influenced by social factors. In his study of physics and psychics in late Victorian England, Wynne shows how "the social contexts of use" influenced the development of late Victorian physics.

Woolgar's analysis of discovery accounts in pulsar research indicates that scientists and the sociologists who study their activities interpret discovery accounts in different ways. Some will, for example, see the accounts as consistent while others will see them as contradictory. And Forman defends in persuasive detail the argument that German physicists in the Weimar period accommodated their science to a hostile sociocultural milieu.[5]

In order to acquaint the reader with the concept of physics as a social fact more fully, I will discuss two studies that deal with the sociology of physics, and one area of research that focuses more generally on *science* as a social fact: Paul Forman's study of physics in Weimar Germany; William Harvey's case-study of experimental quantum mechanics; and the "laboratory life" studies.

PHYSICS AS AN ADAPTATION TO THE SOCIOCULTURAL MILIEU

The relevance of Forman's study of Weimar culture, causality, and quantum theory to the present study has been suggested by several of my colleagues. The obvious place to begin is with Forman's own suggestion that there are parallels between the period he studied, 1918–1927, and the contemporary period. Thus, early in his paper, *before* having presented and defended his argument that German physicists accommodated their science to a hostile sociocultural milieu, Forman claims that the *contemporary* relations between science and society involve the accommodation of scientific ideology to a hostile cultural climate. He notes further that the absence of cynicism and calculation in the "responsiveness" of the scientists suggests that they themselves participate in the antiscientific sentiments of their critics.[6]

Forman introduces his study with a suggestive remark by Max Jammer which includes the thesis that external influences prompted physicists to hope for, search for, and embrace an acausal theory. Forman outlines a model for such an ideological accommodation. Where scientists and science enjoy high prestige, they are generally free to ignore external pressures and to respond to internal disciplinary pressures. This generally means they are free to conform to traditional ideologies and conceptual orientations. A loss of prestige requires countermeasures. If altered public values produce a negative image of science, then scientists must try to alter the public image of science so that it is once again in step with the altered values. But this cannot be accomplished without altering the scientific world view; and this may even lead to changes in the very foundations of science.

What, then, was the situation in Germany during the 1920s; what parallels are there to the 1960s and 1970s in the U.S. and Europe? The first parallel is the relationship between physics and mysticism. Max von Laue, for example, took note of the charges brought against natural science by the Rudolf Steiner school. Planck applauded von Laue's counterattack on Steiner, and went on to complain about the widespread antiscientific currents of the time, manifested in such forms as spiritualism, occultism, and theosophy. Ostwald and Sommerfeld criticized a "rampant mysticism". Ostwald recalled a comparable situation that followed Napoleon's defeat. Sommerfeld was asked to contribute an article on astrology to the most prestigious South German monthly, a request he considered a "monstrous anachronism". His observation that more people in Munich probably earn their living from astrology than from astronomy is echoed in L. M. Branscomb's more recent comment in *Science* on the astrology boom: " . . . there are three professional astrologers in this country for every astronomer".[7]

The second parallel is a wholistic orientation. Bergson, Troeltsch, Litt, Driesch and others are cited by Forman in his discussion of the rise of *Lebensphilosophie*, and its association with, in Litt's words, "the victorious breakthrough of 'wholistic' convictions" in neovitalistic biology, Gestalt psychology, and other disciplines. Driesch's philosophy of biology, for example, encompassed vitalism, wholism, and idealism.[8]

The third parallel is the existence of a sense of crisis pervading the critique of determinism, mechanism, causality and positivism, and the articulation of an "acausal" physics.

Forman goes on to document the influence of Spengler's radical relativism on the intellectual world, and on physics in particular; he refers here to the "capitulation to Spenglerism". Spengler predicted that the fate as well as the salvation of natural science was to be a *new mysticism*. One example Forman cites is Richard von Mises' argument for "speculative natural science", especially in relativity and atomic physics. According to von Mises, the new physics had returned to "the question of the old alchemists"; the numerical harmonies and mysteries of atomic physics were Pythagorean and cabbalistic in spirit.[9] The parallel to Capra is clear; von Mises' statements also recall the rise of the Hermetic tradition centuries earlier in Western Europe. One cannot help wondering whether, if the Hermeticists were in some sense harbingers of the Scientific Revolution, the German physicists' "capitulation" anticipates a new scientific revolution, one which is closer to realization today following the developments in physics between 1930 and 1970. I will say more about this later.

In Section I, I briefly noted the ideological function of Capra's arguments as a justification of physics in the face of counter-culture criticism, and as a reconciliation of physics with the mystical predilections of the critics. There are parallels to this in Weimar culture. Sommerfeld, for example, tried to present an image of physics that would be agreeable to antirationalist and antipositivist critics, and that would raise the prestige of physics in that audience.[10]

There are no internal developments in physics at this time, according to Forman, that can account for the "wave of conversions to acausality".[11] These conversions were quasi-moral, quasi-religious "re-births". As the turn from an acausal philosophy to an acausal practice gathered momentum in physics during the late 1920s, Wilhelm Wien held fast as "the champion of causality". Wien criticized Sommerfeld's "Atomystik". Mysticism, Wien argued, has its place in life; but a physics "in which mysticism governs, or even collaborates, relinquishes the ground from which it draws its strength, and ceases to deserve its name".[12]

Forman concludes his study by citing James Gardner Murphy's remarks to Einstein in 1932: the German scientists " . . . cannot escape the influence of the *milieu* in which they live. And that *milieu* at the present time is characterized largely by a struggle to get rid of the causal chain in which the world has entangled itself". This is, Forman notes, a paraphrase of Schrodinger, who earlier that year had delivered a lecture titled "Is Natural Science Conditioned by the Milieu"?

Two further aspects of Forman's study require attention. One is his note that "Bertalanffy's equation of the renunciation of causality with mysticism is not wholly unjustified; Bertalanffy was here affirming Spengler's view". The second is his finding that the advocates of acausality tended to have "either distinctly progressive political views by the standards of their social class and the German academic world, and/or had an unusually close interest in, or contact with, contemporary literature":[13]

> On the other hand, with the notable exception of Einstein, those who defended causality tended to be highly principled political conservatives and/or interested in classical literature.

Forman's study illustrates how physics can be studied in relation to its sociocultural milieu. Physics (and more generally, science) can also be studied as a social activity designed to evaluate knowledge-claims and establish scientific or objective facts. Bill Harvey has examined knowledge-claims in physics in a study of a series of experiments designed to test quantum mechanics theory.[14] It is based primarily on interviews conducted between 1976 and

1978 with the physicists involved in the experiments. The experiments were performed during the late 1960s and the 1970s. Harvey shows how the physicists' evaluations of theoretical and empirical knowledge-claims were "heavily dependent on the cultural context" in which they were made. In particular, he draws attention to how the physicists quite self-consciously used non-empirical criteria with confidence in evaluating knowledge-claims.

Harvey refers to the *plausibility* of ideas rather than their *truth content*; he argues that

> . . . a belief that an idea is true or correct presumably means that it also seems plausible to the believer. The usage has the added virtue that it is consistent with a relativist approach; a relativist would argue that a dominant idea need not be true, but need only be perceived (for whatever reason) as more plausible than any rival idea.

The experiments Harvey examined involved testing John Bell's *local hidden variable theory*, which had been proposed as an alternative to orthodox *non-local* quantum mechanics. Harvey identifies three stages in the experimental process: (a) choosing an assumption; this involves examining the sets of assumptions associated with different experimental designs and choosing "the most plausible"; and constructing arguments to "maximize the plausibility of the chosen assumptions"; (b) justifying the chosen assumptions; and (c) dealing with disagreements; this involves establishing the plausibility of experiments using various types of arguments (e.g., inductive, ad hominen, expert, and technical).

Harvey shows that the choice of a high-energy photons design over a low-energy photons design (a) was not made in terms of rigorous arguments but in terms of "reasonableness"; (b) depended on prior commitments to an experimental approach, or the selection of a certain apparatus before or after choosing a set of assumptions. In general, Harvey argues that "the evaluation of rival assumptions was peripheral to the *real* negotiations over gaining access to equipment and funds". Similarly, Harvey shows that justification was also a matter of "reasonableness". Convictions are rooted in a scientific culture that "legitimates and limits the parameters requiring control in the experimental situation, without necessarily formulating, enumerating or understanding them".[15]

In discussing the rejection of Holt's 1973 low-energy experiments, which disagreed with quantum mechanics, Harvey notes that the arguments become increasingly conventional *scientific* arguments in *form* but they do not gain in epistemological substance. Arguments do not suddenly become rigorous; they continue throughout the research process to be based on appeals to what is reasonable and plausible.

Harvey concludes that his study shows that certain beliefs considered well-founded are based on conventional grounds; they are accepted because they are in accord with certain notions current in the culture of physics and in the local culture surrounding work on local hidden variable theory. He offers his results in support of the thesis that "the social negotiation of reality" takes place in a social context. This idea has been discussed in a more strenuous form by "constructivist" students of laboratory life.

The major laboratory life studies focus on the fields of biology and chemistry. In these studies, social scientists develop portraits of laboratory work using standard ethnographic procedures – in-depth interviews; observations; the analysis of notes and conversations, draft manuscripts, and published papers; and (in some cases) direct participation. The laboratory life studies, in spite of different aims, assumptions, and substantive concerns, converge in four basic ways: (a) they stress that inquiry is *constructive* rather than descriptive or cognitive; (b) they reject the distinction between social and cognitive aspects of science; (c) they emphasize that scientific facts are situated *socially*, and are constructed out of the interactions of *contingencies*; and (d) they focus attention on "the discursive accomplishment of scientific objects in scientific practice".[16]

The relationship between documents and scientific practice has been discussed in some detail by sociologists of the laboratory. Latour and Woolgar argue that there is "an essential similarity between the inscription capabilities of apparatus, the manic passion for marking, coding, and filing, and the literary skills of writing, persuasion, and discussion". Scientific writing is "not so much a method of transferring information as a material operation of creating order"; the laboratory is "a system of literary inscription".[17]

Knorr and Knorr have analyzed scientific texts as media for *constructing* reality rather than "data which represent reality". And Zenzen and I have analyzed a set of papers in colloid chemistry as *persuasive* efforts.[18] We do not claim that scientific papers are designed to persuade people that something that doesn't exist does exist; nor that persuasion is the only aim scientists pursue in writing papers. We do claim that scientists need to use a rhetoric of persuasion in order to draw attention to and legitimate their findings. Getting a paper which sets forth claims accepted for publication involves choosing the "right" words, and deciding when and how to use analogies, mathematics, systematic theory, and other resources for communicating research.

One of the major consequences of laboratory life research is that "facticity" has become radically problematic in science studies. This is exhibited in a variety of ways. Zenzen and I, for example, have formulated the

"Rashomon theorem" after the Japanese film in which several accounts of an event are presented:[19]

> We do not claim that our account of work in the laboratory is superior to or more "factual" than other accounts that might be given by other observers (including accounts by the laboratory scientists themselves). *Our* aim is to give *an* account of scientific practice.

The Rashomon theorem states that there are innumerable ways of constructing an account of an event or process; "the status of any given account is not determined a priori by whether it is "true" or "real" in some absolute sense, but by how useful it is in the competitive realm of knowledge production and utilization".

The laboratory studies offer first-hand accounts of how scientists reach consensus on "facts". But the current group of ethnographers of science are not the first students of science to defend the thesis that scientific facts are socially constructed rather than "discovered". In 1935, Ludwik Fleck's study of the history of the Wasserman reaction led him to conclude that facts are grounded in historical contexts and reflect definite styles of thought:[20]

> This is how a *fact* arises. *At first there is a signal of resistance in the chaotic initial thinking, then a definite thought constraint, and finally a form to be directly perceived.*

Fleck's conception of the genesis and development of a fact conforms to the descriptions of fact construction in the laboratory studies:

> In the field of cognition, the *signal of resistance* opposing free, arbitrary thinking is called a *fact*. This notice of resistance merits the adjective "thought collective", because every fact bears three different relations to a thought collective: (1) *Every fact must be in line with the intellectual interests of its thought collective* . . . (2) *The resistance must be effective within the thought collective. It must be brought home to each member as both a thought constraint and a form to be directly experienced*. . . . (3) *The fact must be expressed in the style of the thought collective.*

"Truth", according to Fleck, is neither "subjective" nor "relative", nor simply a matter of "convention". At least, these words are misleading if they obscure truth as "a stylized solution". It is "an event in the history of thought;" truth is "stylized through constraint" in its contemporary context.

Spengler seems to have been the first student of the social roots of knowledge to express the constructivist view of scientific facts in its strong form. Indeed, Spengler introduced this idea at the very beginning of Volume I of *The Decline of the West* in a chapter on the most recalcitrant body of facts, *mathematical facts*. His comments on modern physics as "an immense system of *indices*" and "an expression and actualization of a Culture" anticipate contemporary constructivism:[21]

> Every fact, even the simplest, contains *ab initio* a theory. A fact is a uniquely-occurring impression upon a waking-being, and everything depends on whether that being, the being for whom it occurs or did occur, is or was Classical or Western, Gothic or Baroque.

There are *problems* in all of the studies I have reviewed here. But together they present an image of science – and physics in particular – as a social fact. The critique of parallelism is impotent without the comprehension of physics and mysticism as social phenomena. I have not aimed at deep, critical insights in these last two chapters; but, more modestly, I have sought to bring together some of the notions which give meaning to the idea that physics and mysticism are social constructions. I will shortly make explicit something that I have hinted at throughout the book to this point – that is that ideas are cultural resources and are manipulated in support of individual and collective interests. This applies as much to my ideas as those of the parallelists. That this implies a concept of intellectual struggle as a motive force in inquiry will become clearer as we proceed.

SECTION III

PARALLELISM AND SOCIETY

CHAPTER 5

INTERESTS AND IDEAS: PARALLELISM AS AN INTELLECTUAL STRATEGY

I have discussed physics and mysticism as *adaptations* to local and wider organizational and sociocultural milieux. Obviously, this notion applies to parallelism too. But adaptations are never lonely interactions; they are struggles for power between and among more or less well defined collectivities. In this chapter, I discuss parallelism as a strategy in intellectual confict. Parallelists use the ideas of "modern physics" and "ancient mysticism" as *cultural resources* in their struggle to organize, preserve, and advance their interests (or, in more general terms, "to adapt to" and "to transform" their environs). The ideas of physics and mysticism are used differently by opponents of parallelism, and differently again by people who are not explicitly aligned against the parallelists but who hold opposing views.

Parallelism is a *recurring* strategy in intellectual conflict. I have already noted one historical precedent in my review of Forman's study. In this chapter, I will discuss parallelism in the Renaissance.

IDEAS AS CULTURAL RESOURCES

Social Interests are aims pursued or benefits sought by competing groups in the struggle for material and symbolic resources that group members perceive as (and/or outsiders consider) necessary for group survival, and for gaining advantages in relative power, privilege, and prestige. Barry Barnes and others have helped sensitize sociologists of knowledge and science to the ways in which *ideas* are employed as cultural resources to further individual and collective interests. Ideas, write Barnes and Shapin, "have no inherent properties whatsoever, and have features imputed to them entirely according to their mode of use". This is a "strong" version of the interests thesis. It is possible, however, to conceive of a "weaker" version in which the social conditions under which ideas are originally constructed builds in some interpretive constraints; in this sense, there are no "pure" or "neutral" ideas – all ideas, in one way or another and to different degrees, have built in "predispositions" and constraints which affect imputation. I am sympathetic to Woolgar's argument that sociologists of science should practice "a high degree of methodological reflection, as a first step in understanding the practical management of the many kinds of 'philosophical difficulty' inherent in explanation". He addresses himself in particular to the "interests" thesis. My own aim in drawing on the interests thesis is to suggest an explanatory *perspective*, not to offer a systematic, reflective *explanatory account* of parallelism. In defense of proponents of the interests thesis, it should be noted that "methodological reflection" is often a barrier to conceptual innovation. On the other hand, the lack of reflection Woolgar notes perhaps explains why some sociologists of science have failed to recognize that Marxists have had some experience with an interests thesis and may have some insights (based on practice and reflection) on issues such as those Woolgar raises.[1]

My arguments and rhetoric reflect *my* interests as an advocate of a sociological approach to parallelism. Thus, I enter into the parallelist debate; not, however, as an *alternative* to analyzing parallelism in a "pure" way, but rather by *virtue* of analyzing parallelism. That is, analysis and commitment are – in my view – inseparable. Everything I have said earlier about ideas as adaptations and as social constructions applies reflexively to my study.

PARALLELISM, INTERESTS, AND IDEAS

The "interests thesis" is illustrated by the different ways in which parallelists, anti-parallelists, and non-parallelists treat the ideas of modern physics and ancient mysticism. Interests can be analyzed on many psychological and sociological levels. I will focus on interests expressed in written media.

Parallelists are interested in preserving and advancing an image of science as a liberating mode of thought. For example, Capra, a physicist, presents parallelism as a defense of science against the threat of anti-science movements. He explicitly identifies his aim in *The Tao of Physics* as "improving the image of science" among young people who have turned away from science and toward "Eastern ways of liberation". He also wants to share the "beauty" of physics as he has experienced it.[2]

Capra's aims are echoed in Gary Zukav's *The Dancing Wu Li Masters.*[3] Zukav is not a physicist. He describes himself as a person with "a liberal arts mentality". Zukav claims that he has discovered that physical concepts are fundamentally "quite simple". He listened to physicists and discovered that (a) he could "understand" them, and (b) their discussions sounded like "theological" discussions. His aim is to *translate* quantum physics and relativity theory for people who do not have "a scientific mental set" so that they can "understand the extraordinary process which is occurring in theoretical physics". The "extraordinary process" refers to Capra's thesis that modern physics is converging toward ancient mysticism. Zukav asserts, for example, that "the languages of eastern mystics and western physicists are becoming very similar".

The same physical ideas discussed by Capra and Zukav have been "popularized" without imputing mystical qualities to them. For example, physicist Gerald Feinberg discusses quantum mechanics and relativity theory in non-mystical terms. His audience is not mystically-oriented young people, but the "general public".[4] Feinberg, like Capra and Zukav, is interested in conveying a positive image of science to non-scientists. He wrote his book, *What is the World Made Of?*, he notes, for two reasons. One is that he wants to share the "profoundly beautiful" discoveries of physics with "as many people as possible". Feinberg thus echoes Capra and Zukav. His second aim is to justify the funds that the United States and other governments have provided for physics research over the past thirty years. He explains why the research was undertaken and what the outcomes were on the assumption that the funds were well-spent. But Feinberg does not impute "mystical" features to the ideas of modern physics. Indeed, whereas Capra and Zukav stress the radical incommensurability of modern physics and classical physics (i.e., modern physics breaks with classical physics and returns to ancient mystical ideas), Feinberg underscores the *continuities and overlaps* between modern and classical physics. Capra argues that "paradoxes" are intrinsic to atomic physics, and compares them to Zen koans. One of the major "paradoxes" Capra and Zukav discuss is "wave-particle duality". Feinberg, by contrast, does not mention wave-particle duality, and indeed makes no references to paradoxes in physics.[5]

It is interesting to compare Capra's views on modern physics and ancient mysticism with the views of another spiritually-minded physicist, G. A. Shook.[6] In contrast to Capra, Shook argues that (a) scientists in general do not consider the mystical experience "knowledge", (b) neither scientists nor mystics can give us knowledge of the "underlying reality of nature", (c) "true" mysticism is part of but not the same as religion, and (d) mysticism is elitist and esoteric, and cannot contribute to solving our social problems. All that the mystic "knows", Shook contends, is that he/she has had "an extraordinary experience". Shook is generally critical of mysticism, and makes no claims about mystics anticipating modern physicists. His position reflects his interest in defending the Bahá'í perspective on the central role of the Prophet as the vessel of God's revelations, and the "true religiousness" of Bahá'ís.

The differences between Capra and Feinberg, and Capra and Shook, illustrate the thesis that the attributes of ideas are imputed *to* them and are not inherent *in* them. In the following sections, I want to examine how parallelists *use* the ideas of modern physics and ancient mysticism.

MYSTICISM AND PHYSICS IN PARALLELIST THOUGHT

What do parallelists gain by asserting that mysticism is a homogeneous phenomenon, a way of experiencing reality directly, and an experience that transcends sociological analysis? Capra contends that those young people who are turning to Eastern ways of liberation and whom he wants to reach have a marked anti-scientific attitude. Given his interest in improving the image of science among these young people, the best strategy for accomplishing this would appear to be one that brings science closer to mysticism and simultaneously exempts mysticism from scientific analysis, that is, protects it from the very rationality youth are supposed to be rejecting.

First, Capra divorces modern physics from technology. Technology is the "true" source of the evils which young people have mistakenly blamed on science, and on modern physics in particular. Then, he mystifies physics. These two steps – de-technologizing and mystifying physics – are the basis for Capra's argument that physics "can be a path with a heart, a way to spiritual knowledge and self-realization". Insofar as physics is mystical it is (given Capra's view of mysticism) immune to scientific methods and theories. Capra tries to resolve this paradox by adopting a "complementary" perspective: "science and mysticism are two complementary manifestations of the human mind; of its rational and intuitive faculties". Zukav adopts the same solution when he writes that "the practice of Tantra does not mean the end of rational thought". But if the people Capra and Zukav are trying to communicate with have rejected rational thought or feel alienated from it why should they accept this paradoxical resolution? Perhaps in the end, by titling his last chapter 'The End of Science', Zukav offers those alienated from science and rationality a stronger message than Capra does by arguing for complementarity. But neither Capra nor Zukav is in fact prepared to sacrifice science and rationality in the interest of appealing to anti-rational sentiments. This should not be surprising once we recognize that Capra and Zukav are writing on behalf of "rational science".[7]

What advantages can parallelists gain from portraying modern physics as an essentially "simple" (Zukav) homogeneous phenomenon which can be understood in non-mathematical, non-technical terms, and which is moving unhampered toward the world view of ancient mysticism? Parallelists assume that the simpler, more consistent, more homogeneous, and more stable the ideas of physics are made to appear, the more appealing they will be to people who are disenchanted with science. Moreover, in order to get physics and mysticism to correspond, they seem obliged to sustain the mysteries of mysticism while they simultaneously "simplify" physics. This simplification has to be carried out in such a way that physics can be (paradoxically) communicable to laypeople but "mysterious" (or at least open to "intuitive" understanding). This aspect of parallelism is illustrated by Zukav's remark that physics stripped of mathematics becomes "pure enchantment".

The ideas of modern physics and ancient mysticism are cultural resources which are used in different social contexts to serve different individual and social interests. Parallelists impute meanings to physics and mysticism which in their view will make physics appealing to people who are untutored in science, and to people who, tutored in science or not, have turned away from science toward ancient (and especially Eastern) mysticism. Parallelists are interested in preserving or advancing what they perceive to be the interests of a scientific community threatened by challenges to its integrity, autonomy, and power.

ANTI-PARALLELISM

Contemporary parallelism is a reaction and an alternative to counter-culture critiques of science, as well as to anti-science, anti-religion, and anti-mystical sympathies and movements. It is in turn the target of anti-parallelists. Isaac Asimov's anti-parallelist essays illustrate the competition for public sympathies and resources among parallelists and anti-parallelists.

Asimov argues that the conclusions reached by scientists are based on observation and measurement and cannot be equated with quotations from Eastern, Celtic, African, or Greek folklore, mythology, and philosophy. He specifically opposes Capra's arguments in *The Tao of Physics*.

Asimov's general argument is that there are only three things that may be happening to the universe in the long run: (a) it may be unchanging (i.e., there is no "beginning" and no "end"), (b) it may be changing in one direction (i.e., there is a distinct beginning, and a different ending), or (c) it may be changing cyclicly:[8]

> All the sages who have speculated on the universe intuitively must come up with one of these three alternatives. And, all things being equal, there is a one-in-three chance of their having duplicated whatever conclusions science eventually comes to on the subject.

His reply to Capra is that if the Eastern sages know as much about the universe as physicists do, why not turn to a reading of the Taoist texts to discover the *unanswered* questions of modern physics? Of course, Asimov argues, such a strategy would be futile.[9] But then, if ancient texts can be "properly understood" only after physicists have reached their conclusions, what scientific value do the ancient texts have?

> What nonsense all this supposed intuitional truth is, and how comic is the sight of the genuflections made to it by rational minds who lost their nerve.
>
> No, it isn't really comic; it's tragic. There has been at least one other occasion in history when Greek secular and rational thought bowed to the mystical aspects of Christianity, and what followed was a Dark Age.
>
> We can't afford another.

Asimov defends the idea of science as a rational enterprise, standing alone among all modes of thought on account of its unique methodology. He does not deny the role of *intuition* in science; but the truth or falsity of a conclusion in science must ultimately be decided by *reason* and reason alone. Asimov stresses deduction and induction from observations and measurements, replication, and consensus as the hallmarks of science. He stands opposed not only to parallelism but to all perspectives (including by implication social studies of science) which challenge his conception of science as pure reason and unadulterated objectivity. Thus, he defends the very aspects of science which Capra's "counter-culture youth" have rebelled against.[10]

There is a softer argument against parallelism based on the notion that ancient sciences should be studied in terms of their sociocultural contexts and not in terms of their relationship to the development of modern science. The ancient sciences *might* be relevant in some way to our current situation, but this is a matter for study and not something to be taken for granted. The argument is generally applicable to modes of thought, and allows for the possibility that the original meanings of the mystical traditions *might* be useful in dealing with current issues and problems in science and society.[11]

Stegmüller offers an implicit critique of parallelism in considering whether Newton *himself* could have formulated his theory a thousand years earlier, let alone someone who lived a thousand years before Newton. He argues that given nothing but his genius as a physicist and mathematician, Newton could not have done so. The reason Stegmüller gives is that a theory requires special preliminary work (in this case, the work of Kepler and Galileo, for example), and centuries of transforming qualitative phenomena into quantitative data.

This implies that Newton's theory could not have been arrived at through some sort of revelation that by-passed the groundwork. This argument is vulnerable, however, to the extent that it rests on the notion that knowledge develops linearly and in an invariant sequence.[12]

The defense of science in the face of actual or potential threats to its lofty position in the industrialized and industrializing nation-states is sometimes aligned against anti-science movements *and* the rise of religious consciousness movements. These two movements are linked in the works of Theodore Roszak. His arguments are not always a matter of simply setting up opposing camps. In one place, for example, he argues that the mystic has accepted "the fullness of human experience", and has therefore been "more truly scientific than the conventional scientist". It was science as an intellectual activity, however, that was the object of the counter-culture offensive Roszak helped to articulate. The validation of the improvisations of Kerouac and Ginsberg, irrespective of their artistic value, tells us, according to Roszak, something about a generation seeking creativity in an "art unmediated by intellect".[13]

The contrast between Capra and Roszak illustrates the important point that the critique of mechanistic or materialist science can have at least two consequences. In the case of Capra, the alternative to mechanism is a complementary relation between "new physics" or "wholistic science" and mysticism. For Roszak, the alternative is a "new Gnosis". In both cases, however, ancient wisdom is resurrected.

Finally, there is some evidence of *internal* conflict among the parallelists. W. I. Thompson, for example, believes that physics and mysticism are compatible. But he is suspicious of the type of parallelism being worked out in the collaboration between Gopi Krishna and Carl von Weizsäcker. Thomspon is worried about the emergence of a "knowledge business run by an elite corps of mystics and physicists". He might be equally wary of the somewhat looser relationship between Capra and Pir Vilayat and the working relationship between Bohm and Krishnamurti.[14]

The internal and external conflicts parallelists are engaged in can be understood as part of the struggle between scientists and intellectuals for societal resources. More broadly, they can be viewed as part of the struggle for power among individuals, groups, and classes. The emerging dialogue *between* parallelists and anti-parallelists, and conflicts *among* parallelists reflect a potential for new struggles, and new forms of super- and sub-ordination in the intellectual community and in the wider society.

Parallelism, as I have noted in earlier sections, is not a new intellectual strategy. Not only are there parallels in the relations between physics and society in the 1920s and 1970s; there are also parallels between these two periods and the Renaissance, when we consider science in general. I would, therefore, like to extend this discussion to consider the Renaissance period. I want to note certain intellectual parallels with the present, and take note of their relationship to broader sociocultural events. My objective is to expand the context for viewing physics – and more broadly science – as adaptive. Later in this book, I will refer to these historical parallels in discussing parallelism and society today. Obviously, *my* discussion of parallels should be read with the cautions I outlined in Chapter 2 with respect to physics-mysticism parallelism in mind. Parallels, in the sense of anticipations in or the originality of earlier modes of thought, make sense to the extent that past and present settings and experiences are similar, elicit similar behavior and generate similar symbols. But environmental and sociocultural changes affect the *information* content of symbols quantitatively and qualitatively. Comparisons based on assumptions about the information content of symbols (and, thus, the deeper comprehension they reflect) become increasingly stripped of meaning as societies and consciousness change qualitatively (evolve).

PARALLELIST THOUGHT IN THE RENAISSANCE

Parallels between the present cultural climate and that of earlier periods such as late Egyptian antiquity, the waning years of the Roman Empire, and – notably – the late Renaissance, have been pointed out by many scholars. The late Renaissance has been of particular interest to students of occultism, mysticism, and science in contemporary society.[15] Moreover, some scholars argue that modern science retains fundamental links with or residues from the animistic, spiritualistic physics of the Middle Ages, the Renaissance, and the early Newtonian era.[16] Let me turn, then, to cataloguing some of the characteristics of intellectual activity in the Renaissance and the early Newtonian era which are remindful of contemporary parallelism. These are offered, tentatively, as identifying characteristics of parallelism. Parallelism may, in fact, be a recurrent intellectual strategy. However, I will not argue that "there is nothing new under the sun", nor that intellectual life is simply a "circulation of ideas".

Ancient wisdom was glorified in the works of thinkers such as Pico, Campanella, Henry More, and Roger Bacon. Pico listed Homer, Orpheus, and Zoroaster among the Ancient Masters who possessed exact and absolute knowledge. In 1632, Campanella wrote to Galileo that heliocentricity is "a return to ancient truth" and portends "a new age". And the Cambridge

Platonist More described Cartesian mechanics as a truth known to Moses, preserved in the Cabbalist tradition, and divinely revealed to Descartes.[17]

Roger Bacon's views on "that rare and profound philosopher" Hermes are also parallelist. The "journey eastward" in search of occult wisdom is not only a feature of contemporary parallelism and the late Renaissance; it is a tradition that stretches from Pythagoras to Goethe. It is also interesting to note the early medieval precedent for glorifying ancient wisdom, in particular the "discovery" of Pythagorean and Platonic anticipations in Moses and the Prophets. This was a search for a single tradition running through all earlier thought, a *prisca theologia.*[18]

The route to absolute knowledge was conceived as a mystical path. Marsilio Ficino, for example, argued that Hermes Trismegistus was able to put aside "the fogs of sense and of fancy", and open himself to Pimander, the divine mind. Thus, Hermes Trismegistus was able to contemplate "the order of all things" existing in or flowing from God. Ficino, along with Pico, inspired *a new gnosis* as opposed to a new philosophy. The new gnosis, however, was the *old* gnosis resurrected. Similarly, Roszak argues today for a new gnosis but notes that there is nothing we can add to the old gnosis.[19]

Syncretism is a basic feature of parallelist thought; it implies, in this context, a general "intellectual muddle" in which more or less distinct modes of knowing (e.g., philosophy, magic, religion, mysticism, and science) become inextricably and more or less confusedly engaged with one another. Yates, commenting on Bruno's views about the infinite universe and its innumerable worlds, writes:[20]

> Such concepts are not with him primarily philosophical or scientific thinking, but more in the nature of hieroglyphs of the divine, attempts to figure the infigurable, to be imprinted on memory through imaginative effort to become one with the universe . . .

Closely associated with syncretism is a *failure of criticism* and the subordination of criticism to advocacy. Misdated texts (e.g., the *Corpus Hermeticum* and Pseudo-Dionysius' *Celestial Hierarchies*) play an important role in Renaissance thought. Such things are not prominent today, but they are not unheard of. Today, the failure of criticism is reflected in the tendency among parallelists to defend their positions without considering the pitfalls I sketched in Section I or the sociocultural and historical materials I am discussing here. I will consider later, however, the possibility that a loosening of critical faculties may stimulate human thought at certain junctions in its development. The same may be said of another characteristic of parallelist thought, *the negative way*, or "learned ignorance". From the unknown fourteenth century author of *Cloud of Unknowing* to Nicholas of Cusa we find an appreciation of the negative way. Pseudo-Dionysius' negative theology had a wide appeal as *the* method for approaching the divine. The ebbing of criticism is reflected in the *liberties* that parallelists take with "old wisdom", as I indicated in Section I. In the Renaissance, we find Paracelsus twisting and altering the Hermetic tradition "in strange and original ways"; and Bruno "wandering" through Book III of Cornelius Agrippa's *De occulta philosophia*, picking, altering, and rearranging material to suit his purposes.[21]

In the late Renaissance, as in Weimar Germany, and the United States and elsewhere in the West today, *wholism* is a fundamental theme in religio-mystical, magical, philosophical, and scientific circles. It appears, for example, in the Hermetic-Cabbalist tradition, and is fully developed in Francesca Giorgio's *De Harmonia mundi* (1525) where microcosm and macrocosm are harmoniously united and One. The concept of wholism appears to be more complex in the Renaissance by comparison with the concept of wholism in the Pythagorean medieval tradition, or Cabbalism. It may be that, as I suggested in Section I, a new form of wholism is emerging in which entities are treated *as wholes* and not as wholes made up of interrelated and interdependent parts. This distinction is not always conceptually or terminologically clear; but the former is sometimes referred to as *holism* and the latter as *wholism.*

The borders between disciplines or traditionally distinguished modes of knowing are fluid for parallelists. This can be constructive, and there is a clear and widely recognized value in breaking down disciplinary and other barriers among modes of knowing. It can also be vulgar and dilettantish; parallelists are more likely to cross barriers in this sense. It is not always easy or possible to make this differentiation, but a critical approach can reveal problematic border crossings, bridges, and boundary changes. One must be careful, though, since even the problematic cases may directly or indirectly lead to solutions for problems which have failed to yield to prevailing methods and assumptions. Bruno's interpretation of the Copernican heliocentric diagram as a divine hieroglyph is an example of the fluidity of the borders separating modes of knowing in the Renaissance. So is Bruno's claim that there was a divine quality in Fabrizio Mordente's new compass. To Bruno, the Nolan who knows how to interpret such things, the compass was a "wonderful revelation", something Mordente was too ignorant (he was "unenlightened") to realize. Number symbols were not used mathematically but with what Kepler recognized as "Pythagorean intentions".[22] These instances are not so different in principle from the transformation of bubble chamber photographs into revelations of mystical realities, or the view among some mystics that they are "awake" and others are "asleep".

There is no direct contemporary parallel for the Renaissance emphasis on *mnemonic arts applied to the goal of reflecting the universe in the mind*. However, such exercises as those involving the use of holography analogies, and various monistic views of microcosm and macrocosm *may* be functional equivalents. In any case, it is important to note that the mnemonic arts were also important features of early Eastern thought (and of other early systems of thought), and played a role in generating pseudo-modern large numbers and infinites. The development of these arts in the Renaissance culminated in Bruno's system of "magic memory".[23]

The relationship between *wandering* and mystical experience discussed earlier suggests that the wandering lifestyles of Bruno, Campanella, and others like them in the Renaissance probably played a role in stimulating their mystical-magical world views. Today, we should be alert to the relationship between (a) psychic *and* physical mobility and withdrawals, and (b) mystical vision among parallelists.[24]

Just as today's parallelists are confronted by various types of *anti-* or *counter-parallelists* repudiating occult and mystical knowing, so Renaissance parallelists had an opposition to contend with. The evidence on this is clearer in some cases than others. The Kepler–Fludd debate is among the clearer cases. It arises out of Kepler's distinction between genuine (quantitative) mathematics and the Pythagorean approach to number. Less certain is the matter of whether Galileo's repudiation of Pythagorean numerology was an attempt to dissociate himself from the currents of occult science and mystical rationalism.[25]

Finally, and briefly, there are these additional characteristics of Renaissance parallelism which should be noted: (a) the mystical interpretation of ancient wisdom to reinforce present beliefs (More's Christian interpretation of Platonism is an example); (b) the idea of infinitely expanding toward divinity (a contemporary example is the speculation by Bentov and others that consciousness can be expanded throughout an infinite sphere of information); (c) the development of a world picture that provides and reinforces an image of divinity; and (d) a spiritual retreat from the terrible problems of the everyday world in search of peace and purity.[26]

The spiritualist or mystical tendency is not new in physics. One finds it in Kepler, of course, and later in scientists from Crookes, Oliver Lodge, and James Jeans to Eddington and Milne. Labels for modern physics such as "mystical physics" echo eighteenth century debates about *holy physics* (the concept of *physica sacra* which expressed the "perfect harmony" between Newtonian physics and Genesis), and the numerous systems of *physiktheologie*. The residue of Renaissance parallelism is reflected in Newton's view of himself as a "rediscoverer" rather than an "innovator".[27]

The affinities and oppositions among different modes of knowledge in the Renaissance, in Weimar Germany, and in our own times manifest the diversity and complexity of the relationships which can arise between systems of ideas. A brief note on the diversity and complexity of the relationships proposed for science and religion will help to broaden the context for understanding the bases of parallelist thought.

A NOTE ON THE WAYS OF RELATING SCIENCE AND RELIGION

It seems reasonable to conjecture, based on the "interests thesis", that if an individual has internalized a scientific *and* a religious orientation, he or she will strive to reconcile the two orientations. *Any* field of science can be – and probably will be – reconciled with *any* spiritual or religious orientation. There is some evidence that treatises on the reconciliation of science and religion are written by people who have vested interests in both areas – and, on the other hand, that criticisms of arguments for reconciliation come from people with cognitive commitments in one area as opposed to the other.

On the one hand, one finds the distinction between science and religion drawn so that, for example, an eighteenth century religious man felt disposed to oppose Newton with a treatise titled *Moses' Principia*; this at the same time shows the strong influence of Newton, or the Newtonian world view – especially as a threat or an antagonistic force in the struggle for political and/or intellectual power. On the other hand, we find Hitchcock arguing in the middle part of the nineteenth century that geology in particular and science in general may throw light upon the truths of Scripture. Interestingly, whereas today many people are inclined to see the religious implications of science evident primarily in physics, Hitchcock finds geology the branch of physical science most "prolific in its religious application". As we might expect, there is a wholistic quality in his approach. Lecture XIV in his book is titled 'Scientific Truth, Rightly Understood, is Religion'. Here, Hitchcock spells out his view of the "unity of design linking the whole into one mighty scheme". The title of his book – *The Religion of Geology* – reflects the cultural climate of his time as much, perhaps, as Capra's title, *The Tao of Physics*, reflects that of our own. Hitchcock was Professor of Natural Theology *and* Geology.[28]

Duhem reconciles Roman Catholicism and physics by treating them as mutually exclusive. G. C. Henry, Jr., by contrast, reacts to Christian hostility to science (to the theory of evolution, for example) by arguing that the Judeo-Christian religion is compatible with Western science. Moreover, he views mathematics as a traditional adjunct to Christian theology, and a factor in the synthesis of science and religion. Henry holds an undergraduate degree in physics from Duke University, a B.D. degree from the Candler School of Theology, an M.A. in mathematics from Emory, and a Ph.D. in religion from the Claremont Graduate School.

An English biologist, W. O. Greenwood, writing in the late 1930s, confronted the crisis in Christianity brought on by a mechanistic science that appeared to abolish the need for God. He argued that science is actually one of the methods by which God reveals himself. There is no contradiction, in his view, between the biologist's mechanistic approach to organisms and Christian teaching and belief.

More recently, the Oxford zoologist Alister Hardy, "a convinced Darwinian", reflected on the years of "hard thinking" and "searching of the literature" that satisfied him that "the spiritual side of man could be reconciled with the Darwinian doctrine of natural selection (and more recently with that of the DNA genetic code)". He approvingly quotes John Langdon-Davie:

> The whole history of science has been a direct search for God, deliberate and conscious, until well into the eighteenth century . . .

Hardy writes in the tradition of process philosophy and theology. He cites Durkheim's views on the "collective consciousness" in support of his argument that we need to go beyond historical materialism. But he argues further that this suggests not the influence of Society but the *existence* of a Platonic realm of ideas, a field of human consciousness that is or reveals the presence of God and expresses the religiosity of human beings.

In 1928, the Harvard geologist K. F. Mather argued that there is no longer "a wide chasm between the natural and the supernatural, between the material and the spiritual". He adopted a monistic perspective. Mather distinguishes two fundamental roles in science and religion which alternate in a "recurring cycle": the prophet, who takes leadership in discovery, and the priest, who preserves the knowledge. This is a version of the "pendulum theory" of change, and has some of the flavor of Kuhn's view of scientific change. Finally, I note that Eddington, a devout Quaker, found room for God's spirit in the "epistemological gap" between what physicists know and what they cannot know; and that Robert Millikan found evidence for the existence of God in the "birth cries" of atoms we know as cosmic rays.

Spiro offers a social psychological perspective on the phenomenon of opposing or linking idea-systems:[29]

> Rather than reject a doctrine, their first preference is to modify or reinterpret it in such a way that, although on the one hand it becomes consonant with their own cognitive, perceptual, and motivational structures, on the other hand it is recognizable as normative doctrine. . . . But shifts in religious ideology consist not only in the modification of some normative doctrines, but also in the acceptance of others and the rejection of still others. The selective process underlying this differential acceptance or rejection of a religious (or any other) ideology is determined by the psychological structure of the faithful, and more specifically by their cognitive-perceptual and motivational systems.

This is a version of cognitive dissonance theory, or coherence theory. The interests thesis is the collective counterpart of this social psychological thesis. The two theses can be summarized as follows with regard to defenders and opponents of parallelism: an individual will tend to defend parallelism if (a) he/she is a practicing mystic who becomes interested in physics; the physics will be incorporated into the dominant and encompassing mystical world view; this will tend to occur with any and all intellectual disciplines encountered, but the tendency will be greater the greater the discipline's level of abstraction, degree of mathematization, and degree of uncertainty (uncertainty being higher for new and borderline disciplines, and for those on the edge of fundamental changes); the high prestige of the discipline among the sciences and in the wider society will also foster incorporation; (b) he/she has earned degrees in religion and science; the tendency is greater if he/she is teaching *and* doing research or clinical or applied work in both areas; (c) he/she is a practicing scientist who has a mystical experience; the tendency is greater to the extent that this experience is followed up by systematic training and discipline in mysticism or related fields; an individual will tend to debunk parallelism if: (a) he/she is in a discipline seeking scientific status (e.g., sociology); (b) he/she has a positivistic view of science (including critics of positivism who have either not shed all their own positivism or who have reversed an earlier critical approach, even perhaps anti-positivism, and returned to the fold).

Physics, mysticism, and parallelism are social and historical facts. Contemporary parallelism cannot be adequately understood sociologically independently of the perspective I have sketched above. In the following chapter, I offer some thoughts on contemporary parallelism as a problem in the sociology of science and scientific change.

CHAPTER 6

PARALLELISM, SCIENCE, AND SOCIETY

PARALLELISM AND SCIENTIFIC CHANGE: A HISTORICAL PERSPECTIVE

The Hermetic tradition in the Renaissance and earlier – with its parallelisms – seems to have contributed to the West's Scientific Revolution in the sixteenth and seventeenth centuries in a number of ways. For example, Pythagorean approaches to number were not mathematics proper, but directed attention to number as a key to nature; this appears to have facilitated mathematical thinking. The writings attributed to Hermes Trismegistus are an expression of this movement. Numerological techniques are associated with the mathematical revolution in sixteenth century Europe, notably in the works of Jodocus Clichtovaeus and Cornelius Heinrich Agrippa von Nettesheim.[1]

The appearance of the Magus ideal in the early Renaissance reflects a "retreat from reason" into the occult, and is paralleled by the retreat from medieval scholasticism. In both cases, the retreat seems to have occurred in the face of barriers to scientific development, and to have actually stimulated "genuine science". This could be interpreted as a sociological parallel to the well-known psychological strategy of "sleeping on a problem" and solving it in a flash of intuition or insight upon awakening. The "retreat from reason", then, may reflect in part an inability to carry science forward, or more generally a breakdown in the capacity of "rational thought" to further individual and sociocultural development. Yates, indeed, offers just such an hypothesis to account for the Scientific Revolution.[2] The first phase of the Revolution was rooted, Yates argues, in the idea of an animistic, magically operated universe, the second phase in the idea of a mathematical, mechanically operated universe. Yates asks (somewhat mystically, unfortunately) whether all science is not, after all, a gnosis, "an insight into the nature of the All, which proceeds by successive revelations"? Her two-stage theory is consistent with Needham's hypothesis, cited earlier (in Chapter 3) on the relationship between mystical naturalism, religious mysticism, and scientific change. Both are echoed in Baillet's reference to "that sleep of the senses in which truth is revealed" in his description of Descartes' visionary sleep of November 10, 1619. Yates writes:

The Cartesian mathesis was a vision of genuine mathematics as the clue to the universe and led to the discovery of a genuinely scientific tool for investigations. A transition has been made to an epoch in which what is still a Hermetic, almost a "Rosicrucian", impulse towards the world results in valid scientific intuitions. But may not the intensive Hermetic training of the imagination towards the world have prepared the way for Descartes to cross that inner frontier?

Yates provides a broad developmental or evolutionary context within which to examine the significance of parallelism. Parallelism may reflect socially rooted individual and social interests; and it may be an adaptive strategy in intellectual conflict and in response to wider sociocultural conditions. But it may also be a manifestation of difficulties in applying rational thought to individual and social problems; and in that sense it may reflect – directly or indirectly – a search for new modes of thought, or at least, a rejuvenation of old modes of thought.

A SLEEP OF THE SENSES – A SWING OF THE PENDULUM

Are we experiencing "a sleep of the senses" in reaction to barriers to scientific progress, and frustration with the futility or impotence of "rational thought"? It is certainly easy enough to locate discussions of such barriers and frustrations, focused on issues of closure in professionalization and bureaucratization, "rationalization", and "ecclesiasticism".[3] Critics of science can be found in the United States Congress as well as the counterculture. Kevles' description of the "greening of America" in the 1970s, a "movement" which stressed the existence and value of modes of knowing beyond reason, logic, and research, recalls Yates' description of the "retreat from reason" in the late Renaissance.[4] If the Renaissance Hermeticists were in some sense harbingers of the Scientific Revolution, is it possible that the physicists' "capitulation to Spenglerism" (the phrase is Forman's) in the 1920s and 1930s and parallelism today are harbingers of a new scientific revolution?

Some commentators on "the decline of science" seem to be engaging in a rhetorical exercise; they don't really believe that science is in any immediate danger. This contrasts sharply with the following remarks by Kenneth Boulding, who *seems* to accept the reality and even the inevitability of the decline of modern science:

Science is expanding into what seems like a very large niche in the human nervous system. But it is a fundamental principle of ecology that no niches are infinite, and there must come a time – the shadow of which, like the shadow of the stationary states, may already be falling across us – when the scientific method will have produced all the knowledge of which it is capable. Then science will become a church, concerned mainly with transmitting the treasures of its knowledge unimpaired from one generation to the next.

Boulding's perspective leaves us with the following problem: if modern science (or, the modern, primarily Western version of "rational thought") is moving toward some sort of closure, what steps can be taken to stimulate a new flowering of inquiry? Several more or less compatible proposals have been made for dealing with this problem.[5]

Bohm approaches the problem by opposing the fragmentation of thought and life with a program for "wholeness" in values, thought, and ways of living. Quinney writes about new modes of knowing already emerging in association with "the emergence of a new level of consciousness and existence" which is vitalizing, not supplanting, rationality. This process "expresses the other side of our desire", subordinated, ignored, atrophied in the splendor of the rise of scientific/technological civilization:

Only now are we beginning to recognize the unseen that has been excluded by our rationality.

He criticizes the excesses of positivism's search for "facts" and phenomenology's search for "essences" as orientations to what *is* and not to what *could be*. *Transcendental* knowing focuses our attention on what could be:

A leap into the transcendental consequently requires a change in the style of life that heretofore has aided scientific development. In order to know transcendentally, one has to live his life transcendentally. We can have a vision of what could be only when we allow ourselves to be free. We must make ourselves ready to think extraordinary things.

He illustrates the trap of rationality which has ensnared us by quoting a passage from Hawthorne's 'The Artist of the Beautiful': Owen Warland, a man who wanted to create a beautiful machine,

... had lost his faith in the invisible, and now prided himself, as such unfortunates invariably do, in the wisdom which rejected much that even his eye could see, and trusted confidently in nothing but what his hand could touch. This is the calamity of men whose spiritual part dies out of them and leaves the grosser understanding to assimilate them more and more to the things of which alone it can take cognizance.

Grange says this in another way, in discussing the revival of occultism in contemporary society:

... it is precisely the obviousness of our present technology that is the source of much of the anti-technology rhetoric that now dominates social and cultural commentary.

And so, we should not be surprised that "magic" is in the air – for the fundamental and often ignored meaning of magic is "the absence of obvious technique".[6]

Grange, like Quinney, considers technology an obstruction and sees the need for a new question and a new quest. His orientation to "the meaning of Being" and "a way of Being Human", in the phenomenological tradition, is nothing if not Quinney's "transcendental way of knowing".

If, as Grange and Quinney (as well as many others) argue, the rise of science and technology has led us down a one-dimensional path, then we now need to take account of other dimensions, other paths we have ignored and forgotten, or are yet to discover. Gouldner's conception of objectivity is relevant here, for it entails that the path toward "truth" now requires paying attention to what we have been silent about:[7]

> Objectivity . . . is not neutrality: it is realism concerning our own situation, desires, and interests. Here "realism" means being aware of the continued vulnerability of reason to interest and desire, of the limits that interest and desire impose on rational discourse. Objectivity, then, is a wholeness paradoxically pursued by stressing insistently and *one-sidedly* the repressed and silenced side of things. It is the effort to overcome the varied and changing limitations of persons and groups through the recovery of what their lives have systematically repressed, distorted, and lost.

In this sense, then, the discussion of alternate realities may simply express the "swing of the pendulum" between rationality and irrationality, wholism and atomism, mysticism and science, or some such extremes. In the light of these ideas, it is interesting to consider Horton's comparative studies of Western and African modes of thought.

Horton has identified common *structural* features in Western science and in traditional African thought. According to Horton, the fundamental distinction between Western science and African thought is not, primarily, that between "personal" and "impersonal" modes of thought; rather, it is to be found in the social situations of the two systems of thought. African thought operates in a "closed" social situation; Western science operates in an "open" social situation. In closed systems of thought, a single view is dominant; in open systems, there are many views and people are aware of different and competing world views. Obviously, "open" and "closed" are relative terms; and it is precisely what appear to be movements toward closure in the relatively open Western situation that trouble social critics and commentators on the decline of science.[8]

In a closed situation, words will tend to have magical significance; there will be little reflection about modes of thought; ideas will tend to be bound to occasions; and there will be no segregation of motives in the search for explanation. Divination will play a prominent role in the pursuit of knowledge; and the attitude toward existing concepts and theories will be protective. Experimentation is likely to be rare. People will not readily accept the fact that they may be ignorant of the causes of events that affect them deeply. Notions of chance and probability will not find ready acceptance. The "open" situation is in direct opposition to the "closed" situation in each of the foregoing respects. Mary Douglas adopts essentially the same approach to differentiating thought systems in different cultures, and in the history of science.[9] The criterion is based, she says,

> . . . on the Kantian principle that thought can only advance by freeing itself from the shackles of its own subjective conditions.

The compatibility of the views of Yates, Horton, and Douglas should be clear. The convergence between Yates and Douglas is especially noteworthy because of the generality of their views. I will refer to the thesis stated by Yates above on scientific change, in conjunction with Douglas' version, as the Yates–Douglas hypothesis. The Yates–Douglas hypothesis suggests that contemporary parallelism is a reaction to perceived closure (loss of adaptability) in modern science, or rationality.

Yinger, in his 1977 Presidential address to the American Sociological Association, argued that we are experiencing a planet-wide civilizational transformation. Under such conditions, he said, it is important for sociologists to undertake intensive studies of countercultures. They can be conceived as "art forms"; and from "the sublime to the ugly", as "voices crying in the wilderness" or symptoms of disorder and as "unintended warnings and illustrations" of the future, they "highlight, dramatize and anticipate drastic problems". The link to the theme of this chapter is Yinger's remark that today's countercultures seek truth through mystical insight. As an attack on the intellectual establishment, this is not new. In the fifteenth century, it was Taborites versus Prague University masters; in nineteenth century England, it was sectarians denouncing the clergy's monopoly on truth. Today, science seems to be implicated in all of the difficult problems we are facing. Some people see a need for alternatives to a science-and-technology which has become "too obvious", one-dimensional, and directed toward an apparent evolutionary dead-end. Yinger describes the situation as a "symbol-less" one, a situation that recurs periodically in history. It is no accident that in seeking to adapt to such a situation, some people turn to mysticism.[10]

Tiryakian, like Yinger, emphasizes the importance of countercultures as forces for social change; but he is more forceful in his treatment of esoteric culture "as a source of ideational innovation in Western modernization". "Esoteric" is defined as "those religiophilosophic belief systems which underlie occult techniques and practices: that is, it refers to the more comprehensive cognitive mappings of nature and the cosmos, the epistemological and ontological reflections of ultimate reality, which mappings constitute a stock of knowledge that provides the ground for occult procedure". Tiryakian argues that "esoteric influences in the form of symbolism, imagery, practices, and cosmologies have been in much of the background of the rise of the scientific disciplines". He identifies Paracelsus as the key mediator between esoteric culture and modern science. Tiryakian's general conclusion is that esoteric culture is a "religious retreat". This "withdrawal" – this "sleep of the senses" – is a reaction by individuals and groups to failures in established symbols and cognitive models. Tiryakian joins other prophets of a new cultural synthesis when he describes the occult revival as part of a new international "cultural matrix".[11]

Charles Singer has described scientific episodes in which the "speculative attitude" is productive. Everett Mendelsohn points out that such episodes are associated with the breakdown of traditional patterns of authority, dogma, and consensus. Like Singer, he notes that such periods are important because they allow new areas to be explored; but they can amplify and exaggerate problems and concepts. This suggests the hypothesis that a loosening of canons of rigor is necessary, occasionally, to get science (or, more generally, inquiry) out of the inevitable ruts it falls into. Wallis' (1616–1703) "bold use of infinitesimals" is cited by Bloor as an example of "the value of the decline of rigor".[12]

The Yates–Douglas hypothesis establishes a context within which the interests of parallelists, anti-, and non-parallelists take on a broader significance. Parallelism, like other strategies in intellectual conflict, ebbs and flows as history unfolds. Its contemporary visibility, following Yates and Douglas, reflects the perceived vitality of claims about the loss of adaptive potential in science and rationality. K. Boulding and M. Wartofsky, for example, argue that science and rationality are getting into ecological and evolutionary "ruts".[13] Parallelism is *one* reflection of the attempts underway to answer claims such as those made by Boulding and Wartofsky. These attempts are aimed at reestablishing the potency of reason in human affairs. The other aspect of this picture is that a perceived or actual decline in the power of science and scientists is a stimulus to groups interested in taking over science's niche in society.

If ours is a time, to borrow Langdon Winner's phrase, of "technics-out-of-control", perhaps we should expect a sense of powerlessness to develop and to be reflected in a resurgence of interest in "the irrational".[14] Following Bourguignon's suggestion, we can suppose that all societies must regulate the relations between the rational and the irrational. Where control of technological, economic, and social factors appears to be in hand, we should expect little support for the irrational; where control appears to be waning, we should expect a resurgence of the irrational or what some scholars describe as a "romantic reaction". Perhaps we are witnessing "a swing of the pendulum". Weiman, writing in the mid-1920's, conceived intellectual history as a pendulum swinging between mysticism and scientific method, and going through a "midpoint of contemplation" (contemplation culminating in the "discernment of God"). This sort of imagery is also found in the conception of scientific change as the product of continuing conflict among three fundamental world views: mechanistic/materialistic, organicist, and magic/mystical. In the early seventeenth century, the struggle engaged mechanists, Aristotelians, and alchemists. Later examples of this basic struggle encompass Berkeley's critique of scientific dogmatism, Rosseau's opposition to the Newtonian orthodoxy of d'Alembert and Condorcet, and the criticism of material science from Crookes to Max Born, P. Kapitsa, Capra, and Roszak. Randall Collins and others carry us beyond dualistic categories such as Romantic and Enlightenment, and Romantic and Anti-Romantic. One of the noteworthy features of Brush's study of quantum mechanics is his conception of the pendulum metaphor. He offers a more complex version of the metaphor than other writers by introducing the ideas of "coupled pendula" (in this case, science and culture), and strong and weak coupling. An even more complex treatment of the pendulum metaphor is possible. If we want to use the metaphor to illustrate the emergence of a *new* themata (to use Holton's concept), we can imagine that (a) additional energy has pushed the limits of the pendulum's swing further out, or (b) the pendulum has, due to a more complex energy input, moved out of the plane of thought we are used to and into a new plane, with a different amplitude and a different period.[15]

COGNITIVE POSSIBILITIES: ANALOGIES OR HOMOLOGIES?

Hassan notes that "current analogies between science, culture, and sundry artistic and spiritual phenomena can prove too facile".[16] Yet, he argues, "it is possible to be at once too rigorous and too timorous in exploring the cognitive possibilities and homologies":

An analogy ceases to be undue when it is posited as a starting point for further verification: the problem now consists in reducing the various phenomena (aesthetic and otherwise) to more rigorous *structural models* and to identify in them not analogies, but rather similarities of structure, structural *homologies*.

Schumacher and Anderson argue that there are *structural* parallels between the *Heart Sutra* (one of the principal *Prajnaparamita* sutras in the Mahayana corpus) and Bohm's notion of a new order in physics. The basic Mahayana view is taken to be "representative of mysticism as it is normally understood". This basic view is *Conze's interpretation* of the *Heart Sutra* as presenting "a dialectics of emptiness which unfolds in five stages". The "dialectics" takes us from the world of everyday things and persons (stage zero) through stage one, where the world is replaced by a "dynamic web of dharmas", stage two, where we recognize that the dynamic web of dharmas is fully conditioned, and stage three, where we free ourselves from the conditioned (the stage of Nirvana, the stage of the Unconditioned). The surface level world is left behind.[17]

Stage four, according to Schumacher and Anderson, represents the special contribution of the Mahayanists. They refer to this as the "depth level", where the exclusiveness of the conditioned and the unconditioned, of Samsara and Nirvana, is denied. Stage five is Noble Silence.

This dialectic implies that the "ultimately real" is "undivided". But – and Schumacher and Anderson consider this extremely important – the ultimately real (undivided) is not separate from what is "divided" (the "not ultimately real" of stage four). They quote the principal *Madhyamika* thinker, Nagarjuna:

That state which is the rushing in and out [of existence] when dependent or conditioned –
This [state], when not dependent or not conditioned, is seen to be *nirvana* . . .
There is nothing whatever which differentiates the existence-in-flux (*samsara*) from *nirvana*;
And there is nothing whatever which differentiates *nirvana* from existence-in-flux.

This view "extracted" from the *Heart Sutra* is, according to Schumacher and Anderson, structurally the same as Bohm's view of implicate and explicate order:

Samsara and Nirvana, as they are understood at the depth level, can be correlated to the explicate and implicate orders, respectively, and the experiential relationship between them is identical: thought-forms or theories serve to lift into attention aspects of what is otherwise essentially undivided.

Bohm's concept of the holomovement is, finally, structurally equivalent to the Noble Silence.

Now it is possible that what links the Schumacher–Anderson project, Bohm's wholistic physics, the Heart Sutra, and even Conze's interpretation of the Heart Sutra is a spiritual or religious quest. What is gained by showing that there is a parallel (correlation) – structural or otherwise – between Buddhism and physics? The objective might be to show an *anticipation*: but, as I have indicated true anticipation is effectively ruled out on several grounds. The primary question then becomes: Why does an individual or group want to or need to argue for anticipation? Another objective might be to show that at some point in contemporary thought we are faced with *the same problems* as our ancestors, and can do nothing better than echo their Noble Silence. This does not seem to be a motive in the cases I have examined. Striving for cognitive consistency seems to be the primary motive. Persons with positive experiences in science and mysticism can be expected to treat them as compatible regardless of any "inherent" affinities or oppositions between the two modes of thought. Parallelism (Capra) and the search for structural affinities (Schumacher and Anderson) are the two major strategies for establishing the compatibility of science and mysticism. In this particular case, neither strategy seems capable of generating anything more than vague and sometimes unreasonable arguments that are strongest just as those points where the fuzziness of one mode of thought is matched by a comparable fuzziness in the other.

Unraveling the sorts of questions I have been considering is a precondition for comprehending and playing a part in sociocultural change. We may, as Hassan writes, be "living in liminal times, at the threshold of changes in the human measure more profound than any we can imagine", or "simply living through the ordinary nightmare of history from which there is no awakening". The "liminal times" hypothesis seems to be warranted by the movement toward a global sociocultural system. The development of transportation, communication, and exchange linkages among the peoples and cultures of the world may, in fact, be a stimulus to parallelist thought. It may also be a stimulus for that minority of scholars representing virtually every major discipline who have been developing relational-wholistic perspectives on reality.[18] Relational-wholism *and* parallelism may be inevitable consequences of (a) the cross-cultural mobility of intellectuals (psychic as well as physical),

(b) the degree to which different modes of thought have been precipitated in sociocultural milieux, and (c) the degree to which the different modes have become separated and distinct from one another. One imperative on encountering this "fragmentation" seems to be to bring it all together. On the psychological level, this is paralleled by the treatment of objective and subjective factors as interrelated and interdependent.[19]

Global, regional, and local ecological and environmental problem are taxing conventional problem-solving methods, theories, and perspectives; at the same time, they are eliciting relational-wholistic alternatives. If we are indeed experiencing a swing of the pendulum, it is likely to be a novel one, since we have never had to cope with *global* limits before. Creating a viable world under these conditions may require crossing a threshold. In concluding Section I, I referred to relationalism-wholism as relevant to an emerging threshold human condition. It is certainly possible that some sort of global adaptation will emerge which will sustain the "ordinary nightmare". But a global adaptation which does not significantly diminish the adaptive and evolutionary potential of life and consciousness will require fundamentally new ways of living and modes of thought. Relationalism-wholism appears to be a prime candidate in the search for workable solutions to the great variety of individual, social, and environmental-ecological problems we face today. In the next section, I discuss this "wholistic way".

THE WHOLISTIC WAY

The idea that the universe is a relational-whole has gained increasing attention in this century. Reality has come to be viewed, for example, as a product of the complementary relationship between copying objective reality and subjectively creating its image in the process of cognition; and Cicourel has argued that perception and cognition are *multimodal* – they are simultaneously visual, aural, and emotional. Thus, the different modalities of cognition can be viewed as complementary. One or another modality may have priority at any particular time. Shifts in prevailing modalities may be an "invariant process" variously referred to as "swings of the pendulum" and "thematic dialectics". On the civilizational level, prevailing modalities may be identified as "ages". Bochner, for example, describes the period 1776–1825 as an "age of eclosion", when "most of our present-day knowledge evolved into its main organizational dimensions". This was a period during which the West's Scientific Revolution became consolidated, and modern science was born. Today, Bochner argues, we live in an "age of synthesis", but one which carries out its imperative – the development from eclosion to synthesis – across the full range of human experiences and to the borders of the planetary system. This universal scale suggests a transformation in the human realm which many observers feel will focus on consciousness. Thus, Schilling, a physicist, distinguishes modern science (associated with what Bochner calls the "age of eclosion") and postmodern science, a period in which human consciousness will be transformed on "a grand scale" (the "age of synthesis"). Schilling, in line with the relational trend I have noted, develops a wholistic-relational view on physical reality and God. His view is understandable given the cultural and intellectual climate and the need for individuals to maximize cognitive consistency. But Schilling tends to follow other more clearly mystical parallelists by religifying our ignorance, and thus sustaining the sort of unapproachable mystery that can readily stifle inquiry and lead to a *secular* "profound anxiety".[20]

Jaspers labels the contemporary period of scientific and technological grandeur "neo-axial"; he views it as a period of "neo-spiritualization" comparable in significance to the "axial" period, 800–200 BC. In the earlier period, new levels of awareness and consciousness were consolidated; today, new levels of awareness and consciousness are again emerging. The comparison with the axial period suggests a radical change, something more than dealing with new information, or developing new value orientations. One possible interpretation of Jasper's view is that a supra-individual, a "collective consciousness" in the literal sense, is emerging (evolving). The debate about the reality of a supra-individual or Superorganic entity has subsided in sociology and anthropology; but there is a continuing interest in the possible "existence" of such an entity. For example, T. de Chardin posits a "nöosphere", and Oliver Reiser hypothesizes a "World Sensorium", a "Psychospheric Cortex circuitry". Reiser locates "the mystical experience" in the "co-consciousness" of a "universal consciousness"; this is perhaps the most exotic interpretation of the idea that a major transformation in human consciousness is under way.[21] How are such ideas to be evaluated?

The relational-wholistic (wholism, for short) view of thought, reality, and consciousness is not new, as I have noted. There is a tendency among contemporary advocates of wholism to ignore its recurrence in the history of ideas, and especially its association in recent times with the rise of *Lebensphilosophie*, idealism, and vitalism. Forman's discussion of the relationship between wholism and conservative and reactionary politics, is especially interesting.[21]

The "new physics" of the early twentieth century and neovitalistic biology, growing and developing in Central Europe, influenced General Smuts, who coined the word "holism" in 1926 (from the Greek word for "whole", ὅλος). Rather than opposing holism to mechanism (as was the tendency in the German setting studied by Forman), Smuts conceived holism and mechanism as two aspects of the evolutionary process. The relationship between the idea of the "whole" and the appreciation of the "organic" (Cuvier wrote, for example, that "each organic being constitutes a whole") has been associated with conservative politics from Plato and Aristotle to Hegel. Smuts views *may* be more liberal than those of the German wholists; but in any case Forman's study suggests that wholism can be linked to totalistic political and religious systems. Some ten years after the publication of Smuts' book, W. O. Greenwood associated "holism" with subordination to God. In the political realm, wholism has been associated with subordination to the State, the Idea, and other totalistic abstractions. But wholism is not inherently totalistic. This is clear, for example, from G. W. Morgan's treatment of wholeness as "the never-final, discord-and-risk-embracing, living unity of different approaches to the world and of the many facets of the human being".[23]

It might help, to begin with, to try to distinguish the *relatively* unified, integrated (undifferentiated) nature of thought in early (e.g., subsistence) societies from the self-conscious formulation of explanations or justifications for the view that reality should be conceived of and treated as a whole. The latter might be called *wholism theory*, or wholism *ideology*, the former *phenomenal wholism*. It would also help to distinguish between *totalistic wholism theory* or *ideology*, in which the emphasis is on Unity, Integration, and Closure, and *individuated wholism theory* or *ideology*, in which the emphasis is on diversity and change. Finally, the distinction between *wholes* as systems of parts (wholism) and wholes *as* wholes (holism) noted earlier needs to be given more attention.

Parallelists argue that modern physics has rediscovered the wholistic world view of the ancient mystics. This argument rests on ideas such as the "bootstrap hypothesis" and the "monistic interpretation" of geometrodynamics. More recently, parallelists and other "wholistic thinkers" have taken note of the Bell theorem in quantum mechanics, and Bohm's concept of holonomy. Bell's theorem, formulated within and restricted to hidden-variable theory, states that if there are subquantal structures, they must be "nonlocal". Two events are nonlocal if they occur at distant points in space but are connected at the instant of their occurrence. According to Stapp, Bell's theorem has the following "profound result": "The statistical predictions of quantum theory are incompatible with the principle of local causes"; Furthermore:[24]

> Bell's theorem shows that no theory of reality compatible with quantum theory can allow the spatially separated parts of reality to be independent: these parts must be related some way that goes beyond the familiar idea that causal connections propagate only into the forward light-cone.

Thus, Bell's theorem – superficially at least – seems to support wholistic arguments.

Bohm is a strong proponent of "wholeness" in living and thinking as well as in physics. His ideas are widely cited in parallelist circles. Bohm, however, has a rather guarded view of the relationships between Eastern and Western thought, as well as of wholism. Bohm believes we should be aware of past teachings in the East *and* West, but does not support imitating or conforming to them. Nor does he believe it is possible "to go back to a state of wholeness that may have been present before the split between East and West developed". In general, Bohm argues that relativity *and* quantum theory imply that the universe is an *undivided whole*. This is a basis for Bohm's popularity with parallelists. Now Sarfatti has argued from a similar perspective based on quantum theory considerations as follows:

> The participator in a particular quantum experiment in a physics laboratory can be the experimenter himself, though on the deeper level of quantum interconnectedness it must also include the general range of all living systems. All consciousness systems, independently of their *spatio-temporal* locations, relative to the experimental apparatus, make incoherent contributions to the total nonlocal quantum potential felt by the individual photons or electrons.

Bohm, along with Hiley, considered Sarfatti's proposal and concluded:

> ... at present, it is necessary to resist the temptation to conclude that everything is connected to everything else regardless of space-time separations. The evidence to date indicates that non-local effects arise under very special conditions and any correlations that have been established tend to be broken up rather quickly so that our traditional approach of analysing systems into autonomous systems is, in general, quite valid.

Bohm and Hiley refer, for example, to the correlation of photons *locally* which must occur before non-locality can manifest itself over macroscopic distances as in Wu–Shacknov type gamma-ray polarization correlations, or photon cascades at optical frequencies. This conclusion qualifies, if it does not contradict, Bohm's statements on wholeness elsewhere. If nothing else it serves notice that for Bohm at least there are still some questions about what treating the world wholistically means.[25]

It is important to be cautious about wholistic thought oriented to any sort of "ultimate reality". References to what is "ultimately real" reveal, I think, an underlying – perhaps even subconscious – spiritual search. Secondly, I don't think it makes sense to move from the idea of levels of reality to the notion that one level is "more real" than another; and this applies to the Noble Silence as well as Bohm's holomovement. Thirdly, whatever its pitfalls, the attempt to achieve cognitive consistency may be adaptive because it stimulates integration, synthesis, or pluralistic interdependence among a variety of modes of thought. The rationale for the latter is better grounded than the rationale for parallelism.

SUMMARY AND CONCLUSION

The parallels between modern physics and Eastern mysticism are in part the product of unwarranted comparisons between symbol systems separated in time and space. They reflect a failure (a) to distinguish between *numerological* and *mathematical* motives, systems and activities, and (b) to differentiate *aggregated* complexity from *developmental* (ecological, evolutionary) complexity. Parallelist arguments rest on mystified versions of mysticism, and truncated models of physics (and science) as a social activity and process. This is related to the tendency in parallelism to draw together the mythical worlds of the mystic who experiences "pure consciousness," and the scientist who experiences "pure knowledge" (e.g., Platonic truths). The study of physics and mysticism as social and psychological facts undermines this tendency and weakens the parallelist argument.

Parallelism is a recurring strategy in struggles for power *within* intellectual communities and *between* intellectual and other communities. It is, in this context, an *ideology*, and springs from the following strategies: (a) the Forman hypothesis – scientists will take measures to counter a decline in prestige aimed at altering the public image of science to make it consonant with altered public values; this will alter scientific values, ideologies, and doctrinal foundations; (b) the prevailing and most prestigious modes of knowing in a society will be exploited by advocates of other modes to legitimate and add prestige to their modes; they will exploit the prevailing modes by "adopting" their methods and rhetoric. In the Transcendental Meditation (TM) movement, for example, experiments, modern technologies, statistical methods, and mathematical models are used to "validate" the experience of "pure consciousness". Bahá'ís cite their prophet, Bahá u' lláh: religion is the "science of the love of God". And parallelists (including some theologians) argue for parallels, analogies, convergences, and complementarities between science and mysticism or religion. Another strategy is to somehow subordinate the prevailing mode, as in the subordination of science to Vedic cognition in the TM movement. Studies of these strategies as well as the ideas discussed below will be aided by the development of a social psychology of parallelism aimed at determining the bases for individual reactions to the existence of plural modes of knowing.[26]

On the positive side, parallelism may be (a) a (vulgar) reaction to "the failure of modern science in its claim to universal rationality",[27] and to the more general failure of (rationalization of) rationality, (b) a (vulgar) mode in a "swing of the pendulum" toward "a sleep of the senses", and (c) a (vulgar) resultant of transactions between more or less distinct cultural traditions. The relational-wholistic perspective in parallelism is congruent with a minority orientation to relationalism-wholism evident in virtually the full range of intellectual disciplines. This orientation appears to be a relevant reaction to new information and an emerging awareness about the nature of reality stimulated by ecological and evolutionary challenges to human adaptability. However, the nature and meaning of relational-wholistic thought remains somewhat ambiguous and elusive; and historical analyses indicate that "wholistic" thought has totalitarian as well as liberating humanistic potentials. There is an imperative, then, for carefully examining relational-wholistic thought in a political-economic context, and for exploring the various meanings and functions of relationalism and wholism.

We may be in a "crucial period" (Tiryakian), a "waning period" whose character (and perhaps whose fate) is captured in the title of Michael Harrington's *The Twilight of Capitalism*, and somewhat more forebodingly in the title of Stavrianos' work, *The Promise of the Coming Dark Ages*; "The Promise" rests on the ability of "a sleep of the senses" to generate a revolution in human values, modes of thought, and behavior. Such a revolution requires a certain loosening of canons of logic and rigor, and *in this sense* the lack of rigor in parallelism may contribute to the development of new, adaptive modes of thought and the further evolution of inquiry.[28]

The Yates–Douglas hypothesis does not have to be based on a belief in scientific progress, the subordination of irrational to rational modes of thought, or the monopolization of "truth" and "objectivity" by "scientists". It can be conceived as referring to a general process in which adaptations (strategies) reach their limits as guarantors of survival and development, and new *and* old adaptations are "experimented" with in the search for new adaptations. The experimental period can lead to failure and extinction, or the development of more or less successful strategies. These new strategies will "work" (for the species, or for particular groups and classes) for varying lengths of time. Thus, parallelism can be understood as both a strategy in the struggle for power and privilege among intellectuals, and as an attempt to work out new modes of thinking and behavior in response to problems of sustenance and growth in human communities.

APPENDIX C

THE DIALECTICS OF PHYSICS, KNOWLEDGE, AND LIFE

I have already noted that David Bohm has developed a perspective on physics and reality that parallelists find appealing.[29] Bohm is, like the parallelists, sympathetic to the "liminal times" hypothesis, and to the need to take Eastern *and* Western, past *and* present knowledge systems seriously. And while Bohm is not a parallelist in the sense that Capra is, he has a view of reality that is informed by a mystical vision. In this appendix, I evaluate Bohm's ideas as an alternative to parallelism. In spite of his mysticism, or perhaps spiritualism, Bohm goes well beyond the parallelists because he links physics, psychology, and sociology together in a well-articulated world view.

Bohm's two-part paper on "quantum theory as an indicator of a new order in physics" (1971, 1973) is punctuated by such marks of his specialty as "vectors", "Hilbert space", and "singularities". But what sets Bohm's paper apart from the papers of most physicists is that his physics and mathematics are integrated with elements of psychology and sociology. For example, he argues that original discoveries in science are characterized by the "poetic communication of creative perception of new order". Original discoveries are rare because "we tend to give such high psychological value to familiar ideas". He points out that "a tendency to regard certain basic notions of order as permanent and unchangeable" dominates the history of physics. This tendency is manifested in an emphasis on *accommodating* new observations to basic notions of order. Scientists and students of science assume that the perception of new orders is something that can only happen occasionally, "perhaps in revolutionary periods".

There is nothing "necessary" about what T. S. Kuhn describes as periods of normal and revolutionary science, according to Bohm. Bohm draws on Piaget's conception of *accommodation* and *assimilation* as two complementary aspects of intelligent perception to support his argument for a science in permanent revolution. Fitting, cutting to a pattern, adapting, imitating, and conforming to rules are examples of accommodation. Assimilation refers to digesting, or to understanding; "to assimilate" means "to make into a comprehensible and inseparable whole (which includes oneself)". At this point, *the person* enters the realm of discourse in theoretical physics. This takes on special significance because Bohm argues that assimilation has to be given primacy over accommodation in intelligent perception. Indeed accommodation is important primarily as an aid to assimilation. Sensitivity to and perception of new orders in science (or what Kuhnians might label "new paradigms") should, Bohm urges, be integrated with accommodating activities so that things do not become so confused and chaotic that they can only be set right through the revolutionary destruction of the "old order". Bohm's argument reduces to the claim that *assimilation* can become the normal way of doing scientific research. That is, scientists can be educated in such a way that they will be "continually ready to drop old notions of order in various contexts, which may be broad or narrow, and to perceive new notions that may be relevant in such contexts". Under such conditions, the primary characteristics of scientific research would be "something similar to *artistic perception*".

Bohm argues that relativity theory *and* quantum mechanics imply that the universe is a single indivisible entity. Thinking in terms of wholes and parts, and the *classical* interplay between *analysis* and *synthesis*, are not valid in Bohm's "holonomy physics". Bohm lets the *lens* represent the old order in physics; the *hologram* represents the new order in physics. He hypothesizes that there is an undefinable and immeasurable "undivided wholeness in flowing movement" underlying our experiences. We abstract things forming and dissolving in the universal flux, or "holomovement". The holomovement, according to Bohm, is like the stream of consciousness that is undefinable but prior to definable thoughts and ideas.

The physical universe, physics, thought, and language, and the person and society are all intertwined in Bohm's wholistic world view. Thought and language, and person and society are more implicit than explicit in Bohm's papers on the new order in physics. In his book, *Fragmentation and Wholeness*, the physical universe and physics are more implicit, while thought and language, and person and society are more explicit.

The basic problem that Bohm tackles in *Fragmentation and Wholeness* (1976) is the fragmentation of the world and human beings. Fragmentation is, he argues, unhealthy and illusory. The "growing series of extremely urgent crises" we are faced with in the world today is in essence the result of trying to live according to the illusion that the separate parts of our personalities, our natural environment, art, science, technology, work, and society are "separately existent". The source of this illusion is the "necessary and proper" division and separation of things "so as to reduce . . . problems to manageable proportions". Dividing and separating are convenient and useful, especially in "practical, technical and functional activities". They become problematic when the selves and worlds they lead to come to be viewed as "autonomous and independent of human will and desire".

Fragmentation is continually arising from the habit of assuming that there is a direct correspondence between our thoughts and objective reality. Actually, Bohm argues, the thinking process and the product of that process, that is, the content of thought, are a single indivisible whole. Bohm refers to this argument as an *insight*. Theories are insights. Bohm reminds us that *theory* comes from the Greek word *theoria*, which has the same root as theatre. A theory can thus be considered a way of looking at the world; it is a form of insight, and not a form of knowledge about what the world is "really" like. This insight must be applied reflexively; so when Bohm argues that things are *really* whole and that fragmentation is illusory he is offering an insight that is neither true nor false. Insights are "clear in certain domains, and unclear when extended beyond those domains". In science, new forms of insight are continually developing; these are clear up to a certain point and then become unclear. Theoretical "progress" is thus a series of insights, and not a process in which old theories are falsified and replaced by new theories. There are no absolute truths, and there is no steady series of approximations to "reality".

Bohm does not deny the existence of an objective reality. But he views it as an inexhaustible diversity and multiplicity of reciprocally related "things" (for example, entities, properties, qualities, systems, levels) in becoming. This "infinity of things in becoming" contains an "unlimited number of relatively autonomous and contradictory kinds of motion". Bohm does not use a Marxist rhetoric but there are affinities between his views and those of Engels in *Dialectics of Nature*.

The culminating insight in Bohm's work is that *theories are world views*. They are parts of world views; and the world views the theories are part of are implicit within them. Bohm sees world views as encompassing *all* aspects of our lives – "including nature, ourselves and how we are related to external nature and to each other". It is the function of metaphysics to make the implicit world views of theories explicit. Bohm's remarks on world views apply to science, to theories, and to insights. We need to be able to alter our world views radically if they are going to perform "their proper functions" of helping us to organize our "ever-changing knowledge and experience in a coherent way".

For world views to function "properly", Bohm argues, the "whole field of measure" must "open to original and creative insight, without any fixed limits or barriers . . . ". In discussing "measure", Bohm combines the old meaning of health, "to have everything in a right measure", and the newer meaning which stresses quantitative proportion or numerical ratio. Bohm points out that both meanings for measure imply the notions of "limit" and "boundary". With right measure, "our over-all world views will cease to be rigid, and the whole field of measure will come into harmony, as fragmentation within it comes to an end".

More than half of *Fragmentation and Wholeness* is given over to a discussion of language and thought, and an "experiment" designed to produce a new way of using words consistent with wholistic thinking. In the "rheomode" (from the Greek *rheo*, "to flow"), movement is primary, and this is reflected in the primacy of the verb. For example, "to levate" in the rheomode means "the spontaneous and unrestricted act of lifting into attention of the question of whether this content fits a broader context or not, as well as that of lifting into attention the very function of *calling attention* which is initiated by the very act itself". The verb "re-levate" means "to lift a certain content into attention again, for a particular context, as indicated by thought and language". Truth is no longer discussed in terms of "separate and essentially static fragments":

> Rather, our attention is called to the general act of *verration*, and to its continuation in a particular context as *re-verration* and *irre-verration*.

"Irre-verration" means holding to a truth persistently beyond its proper limit.

Bohm is no stranger to Hegel, Marx, and Engels. Over the past decade or so, however, his work has been influenced in a direct and significant way by his friend Krishnamurti. The physics Bohm has produced in this period has been characterized as "mystical". The stream of mysticism introduces an element of orthodoxy into what is otherwise a heretical view of science and society. Bohm's hidden orthodoxy stands out in bold relief when his work is contrasted with that of the self-styled dadaist and anarchist Paul Feyerabend.

There are obvious affinities between Bohm and Feyerabend. They both view scientific traditions as world views, see the scientist as a whole person who is part of and a reflection of his/her scientific activity, and oppose falsificationism. Both give primacy to artistic perception and insight, and defend the "incompleteness theorem": insights, theories, scientific traditions, or world views are *always* incomplete. In general, Bohm's work suggests that he would affirm Feyerabend's position that the "research programme" *science* should be subsumed under the "research programme" *free society*. Both men are what Feyerabend calls "Protagorean relativists":

Protagorean relativism is *reasonable* because it pays attention to the pluralism of traditions and values. And it is *civilized* for it does not assume that one's own village and the strange customs it contains are the navel of the world.

Feyerabend is at home with chaos: "Without 'chaos' no knowledge", he writes. Bohm, by contrast, seems to be constantly on guard against chaos. His imagery in this respect is instructive: bringing fragmentation to an end is like eliminating the turbulent mass of vortices in a stream; to regard world views as permanently fixed is "like putting a solid rock into a flowing stream; the result is turbulence and chaos". The search for inward and outward harmony seems to be an essential ingredient of Bohm's work. In fact, the concept of the holomovement, "undivided wholeness in flowing movement", may be the latest version of the "God" insight. Feyerabend has put more distance between himself and theology than has Bohm. I am reminded of the badger's plaintive remark to Merlyn the Magician, who has been mercilessly criticizing communism while proclaiming himself to be an up-to-date anarchist: "I thought that communism was supposed to be a step towards anarchy". Merlyn is doubtful. But certainly there is a sense in which Feyerabend "the anarchist" has carried to its extreme the program sketched by Bohm "the marxist". In this respect, I think, Feyerabend – like Merlyn – is a little more up-to-date.

The insights of Bohm and Feyerabend are alternatives to the form of parallelism defended by Capra. They bring us to the threshold of an intellectual strategy that is based on a "liminal times" assumption but seeks to root itself in a consistent sociological materialism. In my concluding chapter in Part I, I outline such a strategy.

SECTION IV

EMANCIPATORY EPISTEMOLOGY

CHAPTER 7

EPISTEMIC STRATEGIES, SOCIETY, AND SOCIAL CHANGE

INTRODUCTION

The scene shifts in this chapter from my peripheral and episodic interest in physics-mysticism parallelism to my central interest in theory and practice for an emancipatory system of knowledge and culture. I move from a perspective on physics, mysticism, and parallelism as social facts to a general interpretation of epistemic strategies as social constructions. Then I discuss the contemporary epistemic climate, and conclude by sketching an emancipatory epistemic strategy.

I agree with the parallelists, in some sense, when they argue that these are liminal times. It may be that we are experiencing a "swing of the pendulum", a cyclical reaction to the fragmentation and rigidification of epistemic and other cultural strategies. It is more certain that we are developing a critical comprehension of epistemic strategies that can foster our development as *agents* of epistemic and social change. My aim in this chapter, then, is to move beyond my critique of parallelism to an alternative epistemic strategy. I have introduced some of the elements of such a strategy explicitly in, for example, my critical review of Bohm's wholistic world view and my discussion of the Yates–Douglas hypothesis, and implicitly in my critical analysis of parallelism. I will now try to provide a clearer picture of that strategy. I cannot pretend to be in a position to offer a fully articulated, systematic, and aesthetically pleasing picture. This is not merely a matter of my own limited grasp of the crucial issues involved in the project of constructing a sociologically-informed emancipatory epistemic strategy; it is also a reflection of the current state of research and theory in the sociology of knowledge, and the formidable problems that one encounters in trying to develop a coherent integration of the aims and procedures of inquiry and the aims and procedures of a political agenda oriented to improving the conditions under which human beings live and carry out inquiry. Nonetheless, the effort must be made, just as the actions we undertake in everyday life must be carried out in the face of the ignorance, paradoxes, contradictions, and compromises that obscure, harass, and destroy emancipatory motives, aims, and ways of life.

EPISTEMIC STRATEGIES

Parallelism, physics, mysticism, and sociology are epistemic strategies. Human beings are epistemic agents, or inquiring organisms capable of self-reflection.[1] Following Clifford Hooker, an epistemic agent's current states can be considered to be at least partial functions of its past states and current environment. But if our past states and current environments were the only determinants of our current epistemic states we would be what Hooker focuses on, epistemic *engines*. I assume that humans have a capacity for *agency* that varies but nonetheless sets them qualitatively apart from epistemic engines. In contrast to an *engine*, an *agent's* current states are at least partial functions of imagined states, and of preferred, probable, and possible future states, as well as past states and current environments.

All human activities can be conceived as strategies rooted in struggles for survival, power, and privilege. Epistemic strategies rely on more or less systematic knowledge as a resource for solving problems of sustenance and development in individuals, communities, and cultures. Whatever the narrow interests guiding any given epistemic strategy, deeper and broader interests related to the identification and actualization of the potentials in individuals and groups for creative, critical activities are also at stake. And all epistemic strategies are competitors in the struggle for survival, development, and evolution in cultures.

There are numerous hidden and uncritically accepted assumptions that underlie epistemic strategies. These include assumptions about human nature and the natural world, and legitimate ways of experiencing and comprehending reality. Epistemic strategies are developed, utilized, and changed with scarce regard for what they imply about how we come "to know", what we do with "knowledge", the social relations of knowledge, and questions of values and ethics. It is a small step from the idea of hidden or taken-for-granted assumptions to the idea that epistemic strategies are imbedded in and reflect world views. World views are rooted in historical and cultural experiences, and are internalized and modified by invididuals. For any given individual or culture, "world view" will be (a) more or less restricted in terms of how much of reality "world" encompasses, (b) articulated to different degrees, and (c) retrievable through analysis to different degrees.

To begin with, we can follow Scheler in referring to world view as "everything that is . . . accepted as 'given' *without question*, as well as every object and content of meaning in the structural forms of that 'given' without special spontaneous acts, which is . . . generally held and felt as something *that* cannot and need not be justified". What Scheler asserts about "natural world views" can be asserted here, with the understanding that I intend world view to encompass social life, consciousness, and physical reality:[2]

> ... *there is no* one constant natural world view of "man" and ... the diversity of world views reaches into the *categorial structures of the given itself.*

Once we undertake and develop the tools for uncovering and analyzing world views, world views in general (our own included) can no longer be conceived as taken-for-granted. World view analysis depends on a reflexive comprehension of the analyst's world view. Furthermore, world view analysis creates the conditions for competition among more or less well-articulated world views, and for the self-conscious transformation of world views (imaginatively through combinatory and synthetic acts, and by transforming the social structures that generate world views).

Hooker has shown that philosophies of science have "to be construed as offering complete, systematic world views if the arguments between them [are] to be perceptively understood and intelligently evaluated".[3] Completeness and systematization must be understood analytically, since for any given individual or culture world view will vary in scope, explicitness, capacity for articulation, accessibility to self and others, and consistency. Hooker's approach can be more generally applied to epistemic strategies. I have already to some extent carried out a world view analysis of parallelism, physics, and mysticism. My present objective is to sketch an alternative to parallelism *and* to Bohm's wholistic epistemic strategy, guided by the sort of world view approach Hooker recommends. I begin by outlining the social nature of epistemic strategies.

EPISTEMIC STRATEGIES AS SOCIAL CONSTRUCTS

The classification of things in natural and social realms reflects the ways in which we experience nature and society.[4] To the extent that we experience a primitive natural world untouched and untrammeled by humans, to that extent do we construct primitive natural classifications. The classifications arise as we filter and operate on signals on all levels from the genetic to the psychological and social. Classifications will, from the earliest appearance of humans, reflect projections of what people are aware of and know something about into what is mysterious and unknown. As the world humans are born into and live in becomes more populated by human constructions and transformations, our classifications accordingly come more and more to reflect these constructions and transformations. Predicates and classifications become organized into networks, and as these networks crystallize they become systems of rules and laws, and increasingly (in general) constrain what happens to signals we receive and generate.

Experience, mediated primarily by constitutional features of the human organism modified by socialization in early humans, comes to be mediated through a more and more complex network of rules and laws as culture develops. One can conceive of this classification network as an adaptive mechanism; and in fact like all adaptive mechanisms it tends to get into ruts and to lose its adaptive potential. Barring any reason to suppose that classification systems are different in kind from other adaptive mechanisms, we can assume that they are highly vulnerable to extinction. Memory capacity and other features of human beings and culture may carry some potential for mitigating the iron law of devolution; that is, humans may be able to maintain and even raise their adaptive potential in the process of adapting to and transforming themselves and their environs by virtue of their capacity for dealing with huge and diverse quantities of information. The various sections of a classificatory network are more or less stable and subject to different degrees of negotiation; they can be challenged and changed through various kinds of cooperative and conflictful behavior.

As adaptive mechanisms, then, classificatory systems are never structurally identical to or perfect reflections of a static, unified, and singular "reality-out-there". Classificatory systems are the result of a series of "simplifications"; the human nervous system, and culture, do not stand in a one-to-one relationship with "reality". There *is* "a world" or "a reality". This world is, at the very least in relation to humans, infinitely complex. Placing a human "in the world" allows signals from a pre-existing system to impinge on an organism that can image or otherwise process the signals. But some signals cannot be processed at all.

Genetic, neurological, chemical, electrical, biological, psychological, and sociological systems are singly and in combination filters that select, distort, and exclude signals. Of course, a certain degree of stability in the environs of the creators and users would seem to be a basis for the stability of knowledge systems, even though as Bloor is correct to point out the stability of knowledge systems is the *work* of humans.

How do we account for the stability of classificatory systems, or more generally, epistemic systems? According to Bloor, we do not need to appeal to truth or reality; it is enough to note that classificatory systems are actively protected by "elaborating the network". But individuals and groups protect epistemic strategies because they "work". Now they can work in two important ways. They can work in the sense that they help us get along in our daily transactions with the natural and social worlds. In that case, we may find ourselves thinking in terms of natural and social regularities, patterns, laws, and theories. But they may also work in a narrower sense – they may, for example, be psychologically comforting. These two types of success are not entirely independent, but they can be independent enough so that a strategy that is perceived to be succeeding in one realm may be viewed as failing in the other. I do not want to overlook Bloor's main point on this. When we elaborate knowledge networks and protect certain laws, we do not simply protect "our hard won ability to cope with our environment". For this is an activity that must be sustained by social processes. Decisions must *constantly* be made about classificatory boundaries and retaining or discarding laws in the light of experiential conflicts or rival interpretations. Now the social use of nature helps to account for the structural similarities between knowledge and society. As structures of interest vary, patterns of social relationships vary, and so do patterns of knowledge. In order to generalize this theory, we must be able to identify a small number of recurring types of social structures. As Bloor notes, Mary Douglas is developing such a theory. Simply put, what Douglas has done is set up a coordinate system for classifying classificatory systems.[5] Such systems can vary along two dimensions. On the "grid" dimension, they vary from private systems to shared systems. On the "group" dimension, they vary from systems sustained by ego's controlling influence on others to other people's controlling influences on ego. Bloor then asks: how can we relate grid/group to social role?

> Can there really be a unique grid-group rating for an individual as he moves from role to role?

Different methodologies are adopted under different circumstances according to " the varying demands of expediency". In order to expand the theory to deal with this problem, we can turn to constructivism.

The recent work on the social construction of scientific work that I reviewed earlier can be generalized to epistemic strategies. Epistemic strategies are *constructive* (rather than purely cognitive or descriptive), the social settings and contingencies of epistemic work are *embodied* in epistemic products, and epistemic products are *discursive accomplishments.* All epistemic work involves the construction and preparation of tools and materials.

Obviously, the nature of this constructive work varies with the level of technology; this gives rise, for example, to the notion that as scientific work advances, the reality upon which scientists operate comes to be increasingly *artifactual*. Selection processes are found at all stages and levels of epistemic work; it is *decision-impregnated*. Selection implies the existence of a pool of *alternatives*; the complexity and diversity of the cultural and natural contexts of epistemic work will determine the nature and number of alternatives. Epistemic work is also *transformational*. In science, for example, as Latour and Woolgar point out, "facts are constructed through operations designed to effect the dropping of modalities which qualify a given . . . statement".[6] *Social and cognitive aspects of epistemic work are inseparable*. And finally, epistemic objects are *socially* accomplished in particular *contexts.*[7] These basic elements of constructivism are embodied in the modes of reasoning that constitute epistemic strategies. The referent in the following paradigm is "science" as it is generally understood in orthodox history, philosophy, and sociology of science.

KNORR-CETINA'S CONSTRUCTIVIST PARADIGM: MODES OF REASONING[8]

(1) *Practical reasoning*. There are no rationalities that are unique to science. We must ask about "the *savage meaning*" of science if we are going to understand how and why scientists behave the way they do. Scientists rely on common sense rationalities; they are concerned with making things comparable, establishing a "good fit" between observations and interpretations, and timing, predictability, and correct procedures. They are interested in what has worked in the past, what the alternatives are and what their consequences might be, planning strategies, and choices and the grounds on which to make choices. (a) *Cognitive and practical reasoning*. Cognitive strategies are political strategies; scientific choices are investment strategies for maximizing authority and recognition. (b) The distinction between social and scientific "is used to separate the bias, fraud or distortion arising out of social influences, from what is objective or true and has cognitive roots". It is used this way by students of science and by scientists themselves. This implies that the social/cognitive dichotomy must be considered primarily as a resource for interaction strategies. (c) The social/cognitive distinction prevents social scientists from examining its role in scientific practice. (This seems, then, to be at root a distinction between the profane and the sacred). (d) The working

scientist resists being split into social and technical personalities. Practical reasoning is realized in "the *practices* of knowledge production".

(2) *Indexical reasoning*. Indexicality refers to "the situational contingency and contextual location of scientific action", and points to an opportunistic logic.

(3) *Analogical reasoning*. (a) Analogy and metaphor are routine features of scientific and everyday reasoning. They occur in inquiries that lead down "blind alleys", or to "degenerative problem-shifts" and failures as well as "successful" inquiries. (b) "Analogical reasoning is based on a logic of resemblance"; similarity is conceived to be "logically basic", a presupposition in language learning, and independent of analytic criteria. Metaphoric classification is distinguished from other modes of analogical reasoning "by the degree of *distance* or independence between the two conceptual systems brought together through similarity classification". (c) Analogical reasoning is a conservative strategy. It starts from an unrealized solution rather than an open-ended problem. Risks and uncertainties are subordinated to ideas which promise success because they have worked in the past, even if in a different context.

(4) *Socially situated reasoning*. (a) Scientific discourse "refers us to networks of symbolic relationships which in principle go beyond the boundaries of a scientific community or scientific field, however broadly defined" (that is, it refers us to "variable transcientific fields"). (b) Transcientific fields are the locus of struggles for power over "resource-relationships".

(5) *Literary reasoning*. (a) The scientist "has a repertoire of techniques for persuasion", and uses "rhetorical strategies of objectification". Literary resources are used to effect "conversion as well as perversion" in moving from laboratory work to written communications.

(6) *Symbolic reasoning*. Selections "are sustained by the interpretations which constitute both the living and the frozen discourse (in the scriptures and instruments) of an area".

I propose that the modes of reasoning paradigm applies to *all* epistemic strategies. That is, epistemic agents are practical, indexical, analogical, socially situated, literary, and symbolic reasoners.

The epistemic workplace breeds idiosyncratic interpretations of general rules; these combine with local opportunities (available resources) to create idiosyncratic products[9]. Alternatives emerge and decay in a contingency-space created by the slack in the production system. This slack is manifested in the negotiations over and manipulations of resources and interests, or in "mobilization and subversion strategies". Productions "bear the mark of this disorder"; they are "multifaceted hybrids" created out of resources that assume significance "under a given constellation of the field and the immediate environment". The epistemic agenda involves selecting ingredients and instruments (composition), quantifying, and controlling developments through selections from methodological options.

Epistemic objects, products, or facts are, in short, idiosyncratic hybrids; they are constructed out of rationalities-and-contingencies in a decision-making process that is hermeneutical and agonistic (not dialogic). This process takes place "in the slack of a historical contingency-space".

Epistemic agents "tinker"; they "constantly classify their experience in terms of 'what makes sense', and structure their activities in response to questions of *how to make sense* of their results". Tinkering is opportunistic; in different contexts, tinkering around similar problems leads to different solutions.

EVOLUTION AND TINKERING

The idea that epistemic work is a tinkering process is compatible with the concept of natural selection. Francois Jacob contends that there is no appropriate analogy that links natural selection with any mode of human behavior; nonetheless, he notes that it is possible to play with a comparison:[10]

> Natural selection ... works like a tinkerer – a tinkerer who does not know exactly what he is going to produce but uses whatever he finds around him whether it be pieces of string, fragments of wood, or old cardboards: in short it works like a tinkerer who uses everything at his disposal to produce some kind of workable object.

The notion of tinkering must take into account the principle of "limited number of possibilities" (Goldenweiser). The tinkerer principle states that "similar problems tackled by different people in different environments will yield different solutions". But the "limited possibilities" principle points to an extreme where the number of possibilities is one, and an extreme where the number is effectively unlimited; and these extremes are found at every level of culture and consciousness. It is also important to note that there are multiple and overlapping *contexts of adaptation and transformation*. They are not necessarily coordinated or compatible, and their spatial and temporal boundaries vary. Selection processes in these contexts are competitive. Selection criteria on the level of everyday life are necessarily more limited and narrower than criteria on the evolutionary/devolutionary level. Therefore, our choices on the everyday level with respect to selection-and-retention may be at odds with the choices demanded by the requirements for adaptive and transformative activities, and for maintaining or raising the degree of adaptability or transformative potential, in regional and global populations. I imagine history as a reservoir of adaptive and transformative strategies which we can survey and draw from in our problem-solving activities. These strategies are not "things of the past" but "things of the present", not things to *respect*, but things to scan, play with, succeed and fail with, discard, reclaim, transform. We must always ask about the utility and relevance of a strategy, useful or relevant in what sense, in what context, within what spatial and temporal boundaries, for *whom*, with what costs and risks to environments, populations, social classes, and individuals?

THE CONTEMPORARY EPISTEMIC CLIMATE

The transactions between people and their natural, human-made, and human-incorporated environments selectively educate and elicit epistemic strategies. This helps to account for the fact that different epistemic strategies have prevailed at different times in different cultures, as well as in different sub-cultures. The increase in transportation, communication, and exchange linkages among the world's cultures has brought a variety of culturally-specific strategies into contact, and brought the twentieth-century to the edge of global history. This process has also confronted us with new and challenging problems. These problems, from widespread alienation to the deterioration of large-scale ecosystems, demand new epistemic strategies. Capra's parallelism and Bohm's wholism are epistemic strategies which their advocates claim are relevant to the novel problems we are confronted with today. But parallelism possesses several disadvantages as an epistemic strategy designed to address novel, global problems and problems of human growth and development. It is dependent on a naive account of two narrowly and uncritically conceived modes of knowing; it is insufficiently reflexive and so cannot be witness to itself as a recurring epistemic strategy; it operates outside of the sociology of knowledge and so cannot be witness to itself as a social construction and ideology. Bohm's holonomy is a better candidate because it is reflexive. The main disadvantage of holonomy as an epistemic strategy is its stress on harmony and the spiritual quest.

Epistemic strategies take on various disguises (consciously and unconsciously, or subconsciously) to hide theological, apologetic, and ideological quests. It is therefore important to distinguish epistemic strategies in terms of the extent to which advocates have articulated their ideologies, or more generally, their world views, and *then* to distinguish epistemic quests.

When Laplace replied to Napoleon's query about the place of God in *Celestial Mechanics*, "I have no need of this hypothesis", he was not expressing the emancipation of human reason from the idea of God but rather changing the object of worship from God to a lawful and law-giving Nature. Thus Gauss, quoting King Lear, said "Thou, nature, are my goddess, to thy laws my services are bound". Many intellectuals are still carrying out a program of worship in which terms like Reality, Truth, Objectivity, Science, and Mysticism are really substitutes for the term God (or some equivalent); and the belief in and defense of "a real world *out there*" are the last breaths of a moribund religious quest.

Epistemic strategies shorn of theology and apology can be classified as critical or conformist ideologies of knowledge. A critical epistemic strategy is not based on the a priori assumption that "science" or "physics", or "mysticism" is the or a paradigmatic mode of inquiry. It is resistant to the idea that the best way to study science is to use "scientific methods". This resistance is a necessary condition for the *critical* study of science. A conformist epistemology is based on the notion that science is *the* paradigmatic mode of inquiry, and that the best way to study science is to use "scientific methods". The objective of a conformist epistemology is to nourish and faithfully reproduce science-as-it-is with all its reigning social trappings. A critical or emancipatory epistemology creates the possibility for transforming science-as-it-is, that is, science-and-its-social-trappings.

I should stress that I am critical of "science" insofar as that term refers to a specific and delimited epistemic activity. Science in this sense is, along with its social trappings, *one* type of epistemic strategy. It is not, however, monolithic; and it is vulnerable to changes that both further *and* obstruct inquiry. Since epistemic activities take place within transepistemic fields, it is impossible to get an accurate portrait of science by focusing exclusively *on* "science". In another sense, however, "science" can refer to the best possible epistemic strategy; as an "ideal type", it can be used as a synonym for the epistemic strategy I advocate here. When Feyerabend says that the program for a Free Science, or better, Free Inquiry, must be subordinated to the program for a Free Person and a Free Society, I understand him to mean that since it is people individually and collectively who are agents and carriers of knowledge, the quality of their lives will determine and reflect the quality of their knowledge systems.[11]

The Yates–Douglas hypothesis offers a starting point for outlining a theoretical perspective on the contemporary epistemic climate. Let me begin by reiterating the related hypotheses of Yates, Douglas, and Needham:

The Yates hypothesis: the first phase of the Scientific Revolution was rooted in the idea of an animistic, magically operated universe, the second phase in the idea of a mathematical, mechanically operated universe.

The Douglas hypothesis: thought can only advance by freeing itself from the shackles of its own subjective conditions.

The Needham hypothesis: when a certain body of rationalist thought has become irrevocably tied to a rigid and outdated system of society, and has become associated with the social controls and sanctions which it imposes, then mysticism may become revolutionary.

The Douglas hypothesis is the most general of the preceding hypotheses. Let me restate it, at the same time incorporating the ideas of Yates and Needham:

(1) There are, following Horton, two general types of epistemic systems: (a) open, characterized by heterogeneity (different and competing strategies), and (b) closed, characterized by homogeneity (single dominant strategy).

(2) The history of epistemic systems is characterized by "a swing of the pendulum" between open and closed systems.

(3) All epistemic strategies have a tendency to become closed and rigid; thus, all epistemic strategies have a volatile potential for devolution and extinction (I am treating epistemic strategies as adaptations and analogically as species).

(4) Another way of conceiving the polar limits of the pendulum's swing is in terms of "rational" and "non-rational".

(5) The development and evolution of epistemic systems is dependent on conditions and resources which make it possible for a rigidified system to give way to a more open, flexible, complex, and diversified system.

(6) One major resource for transcending an outmoded, rigidified epistemic system is the set of non-rational epistemic strategies. For purposes of clarity, it may be useful to think of this in terms of "orthodox science" (the rational) and "non-science" (the non-rational, the irrational, the a-rational). It might also be helpful to point out that this theory is, so far, quasi-Kuhnian: normal and revolutionary periods follow upon one another, and so long as the process is continuous we can speak in some sense of progress, development, and evolution.

Now under the unqualified guidance of the Yates–Douglas hypothesis, our agency is either non-existent or severely limited; we seem "destined" to undergo periodic shifts between the normal and the revolutionary, the closed and the open, the rational and the mystical, the mechanical and the animistic. But Bohm, as I noted earlier, suggests the possibility of an epistemic strategy in permanent revolution. The epistemic strategy outlined here is conceived in the spirit of Bohm's idea. It is an attempt to answer the evolutionary challenge of designing a system that maintains or even upgrades its adaptive and transformative potentials even while *it* is adapting and transforming. Such a system would contain mechanisms for maintaining and enhancing open-ness. Science has often been ideally portrayed in these terms; but the evolutionary, progressive, or developmental aspects of scientific change have usually been taken for granted. The epistemic strategy I offer can thus be viewed in two ways: (a) it can be viewed as a reaction to the rigidification of science-as-it-is in accordance with the Yates–Douglas hypothesis and the hypothesis of liminal times; or (b) it can be viewed as an attempt (independent of whether or not these are liminal times) to develop an epistemic strategy in permanent revolution, in accordance with Bohm's prescriptive model for science.

SOCIOLOGICAL MATERIALISM

The general epistemic strategy that has guided my critique of parallelism is sociological materialism. I have proceeded under the assumption that all human thoughts are rooted in the transactions between social activities and processes and the physical or natural conditions of life. From this perspective, it is obvious that as (and to the extent that) the social organization of the emerging global culture crystallizes, it will stimulate thoughts about and elicit new epistemic strategies. To the extent that the tensions and potentials associated with the emerging global culture are systematically comparable to earlier and more localized or regional social changes, to that extent we can expect it to elicit well-known epistemic strategies, such as parallelism. To the extent that it poses novel problems for humans, individually and collectively, it can be expected to elicit novel epistemic strategies. Of course, we may not be up to this challenge; we may find that we can only call up old strategies or that what look like novel strategies are not really novel after all. I advocate an epistemic strategy that recognizes consciousness as a complex, multi-layered, multi-dimensional phenomenon but eschews all theological, religious, and spiritual explanations and orientations. This strategy is not rooted in the search for utopia, harmony, or order, it is not based on any claims about ultimate truths, First Philosophy, or ultimate grounds, and it is not clothed in guarantees of any kind. The first goal of the strategy I advocate is improving the conditions under which human beings live individually and collectively. Therefore, it is intimately related to the continuing struggles for social, political, and economic freedoms, and the freedom of individuals from the *authority* of one or a few parts of their selves and from external *authorities*. These struggles I speak of are to remove the various yokes of oppression and repression that burden humanity, and to give people control over their own lives. This is not a program that can be stated in so many words, but an orientation that guides participation in the complexities of everyday life, the life-and-death struggles against annihilative militaristic strategies, and conflictful and cooperative struggles with more compatible epistemic strategies.

As we stand on the edge of global history, there is a clear imperative for a global epistemic strategy. Such a strategy aims to transcend widely accepted cultural (East–West), neurological (Left and Right Brain), epistemological (humanistic-scientific, physics-mysticism), and cognitive (appositional-propositional) distinctions. This is not an innovation but a generalized version of Marx's conception of the negation of science-as-it-is and its fuller realization in and incorporation into "human science".[12] Heuristically, I orient myself to an evolving global culture which cannot settle into any form of ultimate unity. Therefore, the epistemic strategy I advocate must be viewed as a phase in the theoretically endless evolution of epistemic strategies and world views.

Numerous epistemic strategies are emerging which link the plurality of epistemic strategies developed by people at different times and in different places. Wholistic theories have been articulated in a wide range of intellectual arenas. Such theories tend to dissolve the disciplinary boundaries within which they are formulated. From another perspective, wholistic theories may reflect the dissolution of disciplinary boundaries in response to demands for new kinds of information and knowledge being generated by emerging global problems of survival and evolution, and concern for individual growth and development. Wholistic epistemic strategies such as parallelism and holonomy represent more or less narrow personal and collective interests. But there *is* a *global* imperative that is eliciting an awareness of global interests and prompting a conscious search for global solutions. Global interests and solutions are not all compatible, and we can anticipate conflicts among global strategies (e.g., those of multi-national executives versus those of world order decentralist activists). Solutions to global problems will emerge out of negotiations and confrontations between and among groups with different global interests and perspectives. This does not guarantee that every such solution will have the same degree of global relevance; it is even possible that opposing global interests will generate solutions which are detrimental to establishing a viable global culture. Nonetheless, it seems clear that the major arena for working out convergent and divergent interests is the global one.

The epistemic strategy I advocate brings together the variety of cognitive strategies that humans have developed, and transforms them (tentatively, and without pretensions to "universality", or "eternal relevance") into an open, dialectically-changing, evolving mode of *inquiry*. It does this in part on the basis of recognizing the commonalities that all epistemic strategies share. Every epistemic strategy includes every other epistemic strategy to some degree and in some form. Naturally, the possibility of a new strategy does not alter this state of affairs as long as we recognize that *any* new strategy will to some extent carry with it certain earlier forms, that it will make certain forms subordinate, and that over time certain forms will become vestigial and eventually extinct; and new, *effectively* incommensurable strategies may emerge. Furthermore, while sociological materialism is based on the conjecture that all epistemic strategies are social constructions and are constituted of several forms of reasoning, nothing precludes the emergence of new forms of reasoning.

NOTES FOR AN EMANCIPATORY EPISTEMIC STRATEGY

The "system of information", or more broadly, "network of comprehension" (encompassing information, understanding, explanation, knowledge and appreciation) possessed by a given epistemic agent or unit at any given time and classified as "knowledge" is, following Hooker, conjectured to be "foundationless and irrevocably theoretically conjectural (whether linguistically formulated or not)".[13] Epistemic strategies "evolve from ignorance through conjectural creations of conceptual schemas and theories within these . . . ". The scope of an epistemic strategy is not predetermined; at any given time it is at least equal to what can loosely be referred to as the total comprehension that encompasses rational, intuitive, and other levels of knowing in a given epistemic arena. In the process of inquiry, the non-rational levels of comprehension are constantly being systematized, modified, and rationalized. New configurations of comprehension are continually generated as our experiences and the universe change. The domain of rationality is also continually changing and expanding; it is not restricted forever to what we now refer to as "modern science" or to realms accessible to conventional symbols or language. Language, ordinary or formal, is not preferred a priori as the medium of communication and comprehension. And there are no statements or systems in philosophy, meta-philosophy, or meta-inquiry in general which are logically, ontologically, or epistemologically prior to or independent of the theory and practice of inquiry, that is, the actions of humans as epistemic agents. The "best possible" epistemic strategy ideally reflects the full-range of successes and failures in all epistemic strategies past and present. This is an argument for identifying and nourishing diverse epistemic strategies and diversity within epistemic strategies. It is not an argument for parallelism; however, (following Hooker), how epistemic agents and units work, "what is the variety of their possible realizations and which ones are most effective for which purposes (and at which times), is an open question . . . ". It is not, contrary to Hooker, a matter of asking: "Are we really historically in a position to decide between a Zen Buddhist or Tibetan culture and our own – even with respect to exploration of the real nature of the universe"? Rather, it is a matter of being open to adopting epistemic strategies from other times and places, recognizing that in doing so (a) we automatically transform the conceptual fabric of the adopted strategy by removing it from one network of comprehension and fiting it into our own; (b) the epistemic strategy we adopt will have affinities with some portion of our own strategy's comprehension network (as a strategy, not as an anticipation of *our* conjectures). There may be heuristic advantages, at least, in seeking what Hooker refers to as "metacultural rapport" with other epistemic strategies "in the spirit of critical, comparative exploration". The "best possible" epistemic strategy is conjectured to be "a function of a relevant utility ranking and . . . from a collective, historico-cultural point of view it is optimal for the development of human potential to have a range of dominant utilities (e.g., human understanding, technological reliability, technological power, intellectual coherence and unity of understanding, future fecundity, truth, etc.) operative within the culture".

Epistemic activity is the paradigmatic activity of human beings. It is a process of adaptation, exploration, and transformation in which humans pursue survival, physical, mental, and cultural development, wealth and power, and evolution. At the level of everyday epistemic activity, the "best possible" epistemic strategies are characteristically "tinkering" strategies in which local resources and opportunities are chosen and exploited in a "sculpting" of contingencies that results in the production of facts, theories, technologies, and myths. Epistemic activity is constructive, and aimed at "making things work"

The "best possible" epistemic strategy can never be a specific strategy like Western science, Eastern mysticism, quantum physics, or Trobriand Island magic. Demarcationism as practiced by philosophers and others who want to clearly and unequivocally separate science from non-science, rationality from non-rationality, or truth from falsity is not part of the strategy I am outlining. This does not lead us into naive relativism, nor does it eradicate all hope for demarcation. If we examine the widely discussed Azande "poison oracle" case alongside science as some sociologists and historians of science have done, we may readily reach Barnes' conclusion:[14]

> We possess no rationality criteria which universally constrain the operation of human reason, and which also discriminate existing belief systems, or their components into rational and irrational groups. Variability in institutional beliefs cannot be explained by a conception of external causes producing deviations from rationality. Likewise, the culture of natural science cannot be distinctive because of its rationality, in a universal rather than a conventional sense.

But we *can* discriminate (demarcate, perhaps) based on the scale and scope of past successes, and the probability of future successes (on a larger scale and with increasing scope) associated with different strategies. There is a general strategy found in all "successful" knowledge systems – to the degree that the strategy is "successful", to that degree will this general strategy prevail. This general strategy is distinguished by its *capacity* for critical, reflexive, meta-inquiry. This distinguishes the Azande case from at least *some* cases of what scholars have indiscriminately labeled "science".

From a cultural perspective, the best possible epistemic strategy is demarcated from other strategies by the level and degree of development of the *critical* aspects of the various contexts of inquiry, or what Hooker refers to as the scope of the "schema of criticism".[15] Thus, we must be prepared to criticize facts, technological procedures, specific theories, the deepest and most general levels of our conceptual frameworks, the very manner in which we rank these areas of criticism, and finally the established structures of research and criticism themselves. As we move up through these various levels of criticism, the time intervals for exploring and testing criticisms increases; ultimately, this approach can bring a millenia-long cultural and epistemic tradition into question. It is important to recognize that Hooker conceives this critical enterprise as one that must be built into social structures.

Furthermore, each context for inquiry from the organizational to the cultural level must be conceived as an ecology – and so must the whole cultural context be conceived as an ecology. Decision-making takes place throughout a "single, multiply connected net." It is the "total net" that is the significant unit of evolution.

REALISM, RELATIVISM, AND EPISTEMIC ACTION

The sociology of knowledge has not damaged the notion that there is a recalcitrant reality. It is instructive to take note of the following statements by, respectively, Latour and Woolgar, and Knorr regarding science:[16]

> We do not wish to say that facts do not exist nor that there is no such thing as reality. In this simple sense our position is not relativist. Our point is that "out-there-ness" is the *consequence* of scientific work rather than its *cause.*

> A constructivist interpretation of knowledge is not to be confused with an idealist ontology: I do not maintain that reality is produced (constructed) in the sense that its appearance has no independent existence. Rather, this approach claims that once we see scientific products as selectively carved out, transformed and constructed from whatever is, we will also see that there cannot be any warrant in the claim that we have somehow captured (subject to progressive improvement) what is.

With these caveats in mind, let me now outline a perspective on the relativism-realism dilemma.

Spengler was alert to the relativism-realism dilemma prompted by a "naturalistic" interpretation of mathematics. He identified two different kinds of a priori thought-content. First, there are certain things which *seem* to be valid across Cultures, centuries, and individuals. Then there are "necessities of form" characteristic of given Cultures. These two kinds of thought are associated, respectively, with realism and relativism. Spengler argued that there is no way to define a boundary between these two kinds of thought-content; indeed there is no way to even demonstrate the existence of such a boundary. As a consequence we must press our inquires forward without any a priori restrictions based on the demarcation of truths or facts which are and are not independent of human contexts. In other words, since the demarcation problem in this sense is insoluble, we must proceed *as if* no boundaries exist. Specifically, Spengler argues as follows:[17]

> So far, no one has dared to assume that the supposed constant structure of the intellect is an illusion and the history spread out before us contains more than one *style of knowing.* But we must not forget that unanimity about things that have not yet become

problems may just as well imply universal error as universal truth ... Conclusions on the deep and final things are to be reached not by predicating constants but by studying differentiae and developing the *organic logic* of differences.

Wright in his critical analysis of Wittgenstein's remarks on mathematics concludes his discussion of relativism and realism on a note that is harmonious with the views of Spengler, but perhaps clearer as a strategic dictum: we must be prepared to act on the possibility that things might be otherwise – regardless of the strength of our conviction about "necessary truths". In a sense what Wright does (and here I think he clarifies and remains true to Wittgenstein's program) is to show that the relativism-realism dilemma is a *political* dilemma:

... the decision involved when we accept new necessary statements not by way of adopting an explicit convention but as the result of, in the widest sense, proof would be one of *policy*.

The "Cautious Man" according to Wright will grant the acceptability of a *necessary* statement, and grant that "whatever prompts us to accept the statement as necessary – including appeals to imaginability and the like – weigh with him too but the upshot is only that he accepts the statement as "well-founded". This Cautious attitude, more than anything else, is a political attitude – it expresses the "Cautious Man's" *unwillingness* to participate in linguistic institutions such as the institution of proof.[18]

Hooker resolves the relativism-realism dilemma as follows. He admits, to begin with, that some form of skeptical relativism is a reasonable response to the results of research on the social and historical dimensions of science. Such research clearly suggests that science is relative to sociopolitical interests, human values, culture, and history. On the other hand, we experience the recalcitrance of the world daily in the non-random relationships between actions and reactions, acts and consequences. Hooker writes that he sees no better way of dealing with the positive features of relativism and realism than

... to construe cognition as the exploration of possibility structures, to construe theories as hierarchically organized conjectures concerning possibility structures, and to construe systems of norms (expressing values) as conjectural theories.

Like Wright's "Cautious Man", Hooker wants to be a realist, but not the kind of realist who traps him/herself by seeking binding necessities that may prematurely stifle inquiry or close information pathways.[19] Elsewhere, Hooker takes the position that there is no justification for investing a conjecture with positive or absolute belief.[20]

Spengler, Wright (along with Wittgenstein), and Hooker try to avoid the *pathologies* of realism and the *absurdities* of radical relativism. So too Feyerabend separates himself from the radical relativist and the pathological realist by distinguishing between philosophical and Protagorean relativism – and adopting the latter position:[21]

> It is not asserted, for example, that Aristotle is as good as Einstein, it is asserted and argued that "Aristotle is true" is a judgement that presupposes a certain tradition, it is a *relational* judgement that *may* change when the underlying tradition is changed.

All of our constructions, whether we label them fact, theory, or whatever, and whether they take the form of statements, gestures, or whatever are about the intersection of individuals, groups, cultures, and histories. *Everything*, in this trivial sense, is real and true. But what we usually have in mind when we consider realism is the notion that there is a PHYSICAL or NATURAL reality independent of human beings; Hesse describes the features of naive realism:[22]

> ... there is an external world which can in principle be exhaustively described in scientific language. The scientist, as both observer and language-user, can capture the external facts of the world in propositions that are true if they correspond to the facts and false if they do not. Science is ideally a linguistic system in which true propositions are in one-to-one relation to facts, including facts that are not directly observed because they involve hidden entities or properties, or past events or far distant events. These hidden events are described in theories, and theories can be inferred from observation, that is, the hidden explanatory mechanism of the world can be discovered from what is open to observation. Man as scientist is regarded as standing apart from the world and able to experiment and theorize about it objectively and dispassionately.

This view is no longer widely adhered to. Nonetheless, there *are* realists and relativists. What are the implications of their positions? Does adopting a relativist or realist position make it more or less likely that our epistemic activities will improve our ability to solve problems of survival and of individual and cultural development? From this perspective, there is some advantage to a social system – or an epistemic community – that tolerates, indeed encourages, a variety of realists and relativists; and, furthermore, encourages dialogue, competition, tension, and conflict in a generally *cooperative* atmosphere where annihilation, condemnation, terrorism, and bloodshed are minimized if not eradicated.

INTERESTS AND OBJECTIVITY

All knowledge claims are socially constructed and rooted in social interests.

But this conjecture should not be construed as support for naive relativism. Some epistemic strategies are better or more successful than others; some have greater potential for application outside of the regions in which they were originally developed. In general, the narrower the range of interests guiding a given epistemic strategy, the narrower will be its range of applicability. Strategies that reflect a broader range of interests qualify as "more objective". These conjectures require some elaboration and qualification.

The sociology of objectivity is the study of the social and cultural conditions of inquiry and how these conditions affect our individual and collective abilities to construct objective statements and develop objective knowledge. There is a need to conceptualize objectivity in a way that (a) avoids absolutism and relativism, and (b) links commitment to truths (substantive, methodological, and theoretical) with perpetual openness to signals that may alter that commitment (which includes the possibility of becoming aware of or developing new sensory apparatuses). The notion of ideal or transcendental truths may reflect the relative stability of certain substantive truths; methodological truths may be somewhat less stable but they are not, in general, as unstable as theoretical truths. However, since all three types of truths are interrelated and interdependent, it should be clear, even if somewhat paradoxical, why even the most obvious so-called truths or objective statements must be considered problematic. More generally, it might be useful to distinguish between *informational* and *comprehensional* objectivity. An objective statement might then be defined as one that is consistent with the full range of information possessed by the human species at a given time and "known" to a single ideal intelligence. Information exists to the degree that it is possible for human beings to achieve certain effects by carrying out certain actions in accordance with certain principles and to be able to do so repeatedly. Information tends to accumulate into relatively stable and increasingly universal "bundles" which can be pressed into operation in appropriate circumstances (e.g., the objectivity of selecting and eating edible mushrooms). At this level, objectivity approaches limits defined by the boundaries of information niches. But the ability to be objective about mushrooms is not simply a matter of the "facts" about mushrooms; the ability is dependent on intersubjectivity and consensus probabilities.

Objectivity has a second dimension that I refer to as comprehension. It is in this sense that the concept of objectivity as a process is most clearly established. For we can learn to do things, and once we learn to do them, repeat them over and over, generation after generation, with the same degree of success: in this sense, we can say that we "know" something. But our comprehension of what we do can and does change; our theories change. They change because of new information in the form of new signals, new configurations of old information, or new sensory apparatuses which tune us in to new realms of information. Our theories become new world views, so that so-called "eternal" truths change in terms of how we "see" or comprehend them.

The degree of objectivity (objective content) in any given statement is determined by ascertaining the scope and depth of the information available to, or "in", a given individual, group, or community. Thus, the most objective statements that can be made at present about any human experience, can be made by the people, groups, and communities who, or which, have had the widest psychic and physical exposure to human experience as a whole in the present, and in the past through exposure to historical accounts. The most objective statement possible about the nature of, for example, a star (the statement with the highest possible objective content) is constrained by the range of experiences it is possible to have as a human being on this planet. The most objective mode of inquiry at any given time is the mode of inquiry practiced by those who have access to the widest possible range of human experiences, and who are oriented to exploring the limits of those experiences in order to expand their range. Note that while objective statements are referred to an ideal system (the omniscient knower), they are not themselves conceived as "existing" in an ideal or transcendental form or realm.

From an interests perspective, objectivity is more than just a matter of "statements": it is a complex, unfolding process of relationships, feelings, thoughts, intuitions, and "imponderables and ineffables". A sociological theory of objectivity must take into account, for example, the prominence of "unreasonable" modes and motives in the history of science. Discoveries in science can be: (a) the result of "hard work and luck" even when the driving force appears to be mathematical or theoretical "reasonableness" (as in the case of, it now turns out, the discovery of Pluto); (b) stimulated by unusual or bizarre experiences (as in the case of Poisson, whose interest in pendulums may have derived from his childhood experiences of swinging to and fro on a nail upon which he was sometimes hung for safety by a nurse who had to leave him alone for short periods); (c) indirectly by personal traumas (the young Newton, for example, is said to have been stimulated to pursue his studies with greater dedication in order to overtake a boy who stood ahead of him in class – after that boy had kicked him in the stomach); and (d) achieved by the suspension or alteration of conventional canons of logic and rigor (as in Wallis's treatment of infinitesimals). Once we get past positivistic and idealistic reconstructions of science, it becomes easy enough to read the history of science as a record of irrationalities, bizarre incidents, and improbable events. However, that history can also be read as an argument for establishing and sustaining conditions of open-endedness and pluralism.[23]

The sociology of objectivity as I conceive it operates on the assumptions that (a) no objective statement or truth can ever be final or absolute; (b) no system for arriving at truth can be universally valid and unchanging in its foundations; and (c) a broader context for establishing truth always exists than that of any system of knowledge that is given or that is dominant. By definition, the first assumption is not an objective statement or truth; this avoids a classical paradox. By virtue of these assumptions, the sociology of objectivity should contribute to generating new contexts and meanings for truth or objectivity. This perspective, incidentally, tends to dissolve the distinction between gnosio-sociology and the sociological theory of knowledge.[24]

In summary, I suggest that there are *two* types of objectivity: Type A objectivity is generated by persons and groups whose contact with interest arenas is limited but whose focus is phenomena that have a high potential for generating widely applicable knowledge claims. Here, the potential for generalization can be expanded, even though the interest arena remains unchanged in scale and scope. Type B objectivity is generated by persons or groups whose contact with interest arenas is wide; the more the boundaries of these interest arenas are dissolved and the variety of interests homogenized into a general interest, the greater the degree of objectivity the interest arena can generate. The interplay between Types A and B is one form of protection against the maladaptive nature of homogeneous or wildly diversified and independent epistemic strategies.

The *best conjecture* on the nature of reality, following Bohm, is that it is a never ending process of unfolding in which new "things" (hierarchies, relations, wholes) are continually emerging. Reality is complexly determined. It is lawful but indeterminate. Discoverable laws are *indicators* of "mechannisms" which "produce" the phenomenal stabilities, irregularities, and fluxes we experience. It is a multi-level, heterarchical, open system infinitely decomposable (i.e., there is effectively no limit to the ways in which we can "see" the world, trace causal chains, etc.). Laws (or other statements which condense regularities or patterns) indicate mechanisms that operate throughout reality and on every level, but can only be discovered by creating or recognizing approximations to closure in regions of the world. Closures can be approximated experimentally, experientially, or through states of consciousness. Reality is *conditionally* comprehensible because it is continuously changing, even in its essence, and because our closures can only yield stylized indicators of mechanisms underlying our experience. Any given thing is (conditionally) knowable in principle, although it will always have an "incomprehensible fringe" (reflecting the ever-changing unfolding of the universe, as well as the inherent dialectic of things).

Local- or micro-sites can generate statements of wide (macro-) applicability because of the isomorphism which characterizes (and links) all things in the world, and because we *seek* those particular generalizations that indicate mechanisms underlying a wide if not a universal range of phenomena. The capacity of local environments to generate universal statements may reflect the fact that the world is in some sense a hologram (analogically) in which information about the whole is enfolded or encoded (with certain qualifications) in all the parts (however decomposed).

A sociological materialism of epistemic strategies can contribute to identifying the conditions under which epistemic systems undergo closure and stagnate. It can also help to identify the conditions under which epistemic systems resist closure. The literature on science and society illustrates a number of approaches to these tasks. The facilitative relationship between science and democracy has been commented on by Tocqueville, Sigerist, and Merton among others. Merton sees some basis for the provisional assumption that science flourishes in democratic milieux.[25] He notes that science in some form has existed in all kinds of societies; but the crucial question is which societal type(s) facilitate(s) the "fullest measure of development" in science. The emphasis in these and related inquiries is on "external social forces" that facilitate or obstruct scientific activity and progress. Internal social forces that affect science as a social activity, process, organization, or institution are treated incidentally if at all. To fully comprehend science as a social fact, we must attend to social forces on the organizational and institutional levels. Professionalization and bureaucratization are examples of such forces. Both processes have been associated with the emergence of science as a functionally autonomous, progressive social activity. Their continuing impact on science has stimulated some concern among scholars about dysfunctional consequences.

Richard LaPiere argues that the adaptivity of the universities and of the professions has decreased as they have become increasingly bureaucratized:[26]

(1) Rewards for bureaucratic conformity tend to be higher than rewards for innovative behavior;

(2) Bureaucracies tend to exercise some control, directly or indirectly, over nonmembers; thus scientists often have to design their work in accordance with the "rules and prejudices" of bureaucratic organizations; and

(3) Established, or "mature" bureaucracies tend to resist adaptation to changes in external conditions and to resist adopting available innovations.

Bureaucratization, according to LaPiere, tends to subordinate individual to collective decision-making, dividing responsibility for a given decision. This can easily lead to the negation of responsibility, and then to a failure to act effectively with regard to internal organizational problems, or broader "external" societal problems.

The dysfunctions of bureaucratization are reinforced by and reinforce the dysfunctions of professionalization. The two processes are linked at least to the extent that they are concomitant in the modern history of industrializing nations. Professionalization has been associated with the increasing specialization in the division of labour, the knowledge explosion, and the increasing demand for management expertise in highly technical and bureaucratized societies.[27] In the process of professionalization, an occupation becomes "relatively colleague-oriented", with practitioners seeking exclusive rights over naming and judging their mistakes. The goals of professionalization include standardizing, specializing, gaining status for occupational roles and services to society, and "objectivizing", that is, limiting the impact of subjective elements on performance and service. One of the first, and among the foremost, students of professionalization, Carr–Saunders concluded that

> ... taking all in all the growth of professionalism is one of the hopeful features of the time.

The dysfunctions of professionalization, however, arise precisely from the "hopeful" tendency toward occupational demarcation (cf. my discussion of MacKenzie's work at the end of Chapter 13). This creates a volatile potential for subordinating reason to dogma; Freidson writes:

> ... once given its special status, the profession quite naturally forms a perspective of its own, a perspective all the more distorted and narrow by its source in a status answerable to no one but itself.

In his analysis of the medical profession, Freidson argues that while professional autonomy may have facilitated significant increments in knowledge about disease and treatment, it "seems to have impeded the improvement of the social modes of applying that knowledge". Horowitz affirms this aspect of professionalization:

The professional can, by virtue of his professionalism, exempt himself as a scientist from responsibility for the ends to which his scientific findings are put.

The negation of responsibility, as I noted earlier, has also been associated by LaPiere with bureaucratization.[28]

The literature on professionals and complex organizations has traditionally stressed the conflicts inherent in linking the roles "professional" and "bureaucrat" based on differences between "professions" and "bureaucracies". This research focuses on the independent professional's resistance to bureaucratic rules and supervisors, the rejection of bureaucratic standards, and the conditional loyalty to the bureaucracy. But an increasing convergence of bureaucracies and professions has been noted as bureaucrats become professionalized and professionals become bureaucratized. In this convergence, the dysfunctions of the two processes are likely to reinforce one another. Bureaucratization, for example, may reinforce tendencies in professionalization toward occupational closure and dogma with its demands for "reliability of response and strict devotion to regulation". To the extent that the dysfunctions of bureaucratization and professionalization become increasingly salient and converge, we can expect a tendency toward occupational closure, an ethnocentrism of work, and a decrease in the capacity of individuals and organizations to respond to problems in critical and creative ways. Horowitz has remarked that the ultimate consequence of the process may be the "loss of objectivity".[29]

The important point to consider in reviewing the literature on the dysfunctions of professionalization and bureaucratization is not so much what it reflects about particular empirical realities, but the fact that it illustrates the mutability of social facts and the potential that exists in all social phenomena for dysfunctional or pathological transformation. Philosophers and other students of science have acknowledged the potential for evolutionary change in science, but they have not given adequate attention to the potential for devolutionary change inherent in science as a social phenomenon.

Any social structure or function can become pathological. Pathologies may lead to death, irreversible damage, a reduction in vitality, and so on. A "best strategy" epistemic orientation must, therefore, include a theory of pathology and how to deal with it. A preventive approach is inherent in the self-conscious approach to epistemic action, but "doctoring" is not ruled out. Just as there are risks in making decisions at other points in this strategy, there is a risk here of labeling potentially useful changes as pathologies. A risk of the first kind (Type I) involves labelling a potentially useful change a pathology and aborting it; a risk of the second kind (Type II) involves not taking action to treat a pathology because it does not look serious, or because of uncertainty about whether a pathology is present.

A secular, no-warranty evolutionary, emancipatory epistemology recognizes the presence of devolutionary forces and also the long-term tendency of any system to get into a "rut" and to begin to lose its adaptive/transformative potential. One fundamental objective of such an epistemology is to sustain or raise adaptive/transformative potential while actualizing the potential of a given epistemic strategy. This amounts to an experiment designed to determine whether we can "jump" to a new epistemic strategy without in some sense losing local and phylogenetic continuity. This probably is the major evolutionary problem that confronts the human species.

CONCLUSION: CHANGE AND PROGRESS

In order to comprehend reality we must organize in such a way that we maximize actualization of our potential for open-ended inquiry. More generally, if inquiry is conceived to encompass all activities which contribute to human progress and evolution, then the question we must answer is: How must human beings organize in order to adapt, progress, and evolve?

Following Lenski, I define progress as the process by which human beings raise the upper limit of their capacity for perceiving, conceptualizing, accumulating, processing, and utilizing information and energy in the adaptive, transformative, evolutionary process. The relationship between adaptation and evolution is a paradoxical one. On the one hand, survival depends on the capacity to adapt to surroundings, on the other hand, adaptation involves increasing specialization and decreasing evolutionary potential. Adaptation is a dead end. As a given entity adapts to a given set of conditions, it specializes and begins to lose its capacity for adapting to significant changes in those conditions. Sahlins and Service summarize these ideas as follows:[30]

(1) Principle of Stabilization: specific evolution (the increase in adaptive specialization by a given system) is ultimately self-limiting.

(2) General evolution: progressive advance measured in absolute terms rather than in terms of degrees of adaptation in particular environments occurs because of the emergence of new, relatively unspecialized forms.

(3) Law of Evolutionary Potential: "The more specialized and adapted a form in a given evolutionary stage, the smaller is its potential for passing on to the next stage".

If we begin by assuming a Bohmian reality, it is reasonable to consider the hypothesis that Bohmian reality encompasses social reality. Engels outlined such a (dialectical) perspective on physical, natural, and social reality. According to Engels, the laws of dialectics are abstracted from natural and social history. Following Hegel, Engels notes that these laws are simply "the most general laws of the two aspects of historical development as well as thought itself. And indeed they can be reduced in the main to three: the law of the transformation of quantity into quality and vice versa; the law of the interpenetration of opposites; the law of the negation of the negation".[31]

Gurvitch's view of social reality parallels Bohm's view of reality. He paints a picture of social reality in constant motion, filled with tension, "fluctuating renewing, threatened by revolution . . . "; Gurvitch identifies five operational procedures of the dialectic:[32]

(1) Complementarity (contradictory alternatives turn out to be complementary; two polarities are connected by a continuum; polar points pull together, i.e., go in the same direction; and they pull apart, i.e., as a compensatory action).

(2) Mutual implication (things that are heterogeneous or opposite exhibit mutuality and interdependency; they turn out to be immanent, at least partially, in one another).

(3) Ambiguity (ambiguity can eventually lead to ambivalence, and then to polarization).

(4) Polarization ("The tensions between the several factors or aspects [of social reality] are relative; they are observable in different degrees of intensity. Sometimes they reach the proportion of polarization; other times they exist as ambiguities or mutual implication or evidences of complementarity. The paradoxical often resists the dialectical process of polarity").

(5) Reciprocity of perspectives (total identification and separation are denied; mutual immanence, parallelism, and symmetry).

This adds some specificity to my earlier remarks on the dialectics of rational and non-rational modes of knowing.

The present abounds in evolutionary and devolutionary tendencies. To encounter these tendencies without care and passion for one's self and one's community, to meet them technocratically and scientistically, or to sit back and rely on Providential goodwill can only court disaster. A sociological materialism cannot insure that we will continue to do science or inquiry or participate indefinitely in the evolutionary process. It can, however, provide us with a better sense of what has to be done, and what (in terms of available resources) can be done. It can help us identify conditions of evolution and devolution, progression and retrogression. Only our participation in this process of uncertainties can determine whether and to what extent an emancipatory epistemic strategy can take root in this world.

PART II

THE SOCIAL ROOTS OF MATHEMATICS

SECTION V

INTRODUCTION

CHAPTER 8

THE SOCIOLOGY OF MATHEMATICS

The sociology of mathematics has grown slowly on the periphery of developments in mainstream sociology of knowledge and science. Spengler, Hessen, Wittgenstein, Struik, and Needham were among the small group of important contributors to the sociology of mathematics (most of them Marxists) between 1920 and 1970.[1] But their contributions were overshadowed by the development of functional analyses of the social system of science that virtually excluded studies of mathematical activities and knowledge.[2] Mannheim explicitly exempted mathematical knowledge from sociological inquiry.[3] During the last decade, new directions in the sociology of knowledge and science have helped to stimulate interest in and research on mathematics. The renaissance in radical science studies, the "strong program" in the sociology of knowledge, and ethnographic studies of science have been especially important in breaking down the traditional barriers to sociological studies of mathematics.[4] This is a critical development because among all systems of knowledge, mathematics has been the most resistant to sociological analysis and the ultimate arbiter of the limits of the sociology of knowledge.

Mannheim is one of the sociologists of knowledge who denies the *possibility* of a sociology of mathematics:[5]

> Even a god could not formulate a proposition on historical subjects like 2 × 2 = 4, for what is intelligible in history can be formulated only with reference to problems and conceptual constructions which themselves arise in the flux of historical experience.

Similarly, Stark writes: "Surely, there can only be one science of numbers, forever self-identical in its content". Spengler, by contrast, not only affirms the *possibility* of a sociology of mathematical knowledge but defends the extreme position that (a) "*There is not, and cannot be, number as such*. There are several number-worlds as there are several cultures"; and (b) "There is not mathematic, but only mathematics". Spengler carries the historical materialist argument on mathematics to its logical conclusion.[6]

My objective in this study is to outline a sociological materialism of mathematics. This perspective is deeply indebted to Marxist scholarship which has dominated and indeed defined the sociology of mathematics. However, there is a need to go beyond Marxism. For this reason, I consider

Spengler's contributions. There are many things that are problematic about Spengler, not the least of which is a nationalistic political orientation that some liberal and radical intellectuals have referred to as "fascism". Nonetheless, his chapter on numbers in *The Decline of the West* is too provocative and challenging to be ignored. He presents a materialist account of mathematics, on my interpretation, that does not (like Marxism) give up consistency in order to save dogma.

In this introductory chapter, I *briefly* review important contributions to the development of the sociology of mathematics that are outside the mainstream Marxist scholarship that dominates this field.

WITTGENSTEIN

Perhaps the most important thing that Ludwig Wittgenstein has done for the sociology of mathematics is to teach the art of *questioning* what is taken-for-granted and considered "eternal" and "inexorable" in our mathematics. Let us *at least consider*, he urges, the *possibility* that mathematics has a context; thus, it may be that when we say something is "provable", or that it is "true", we must ask " 'provable' in what system", and " 'true' in what system". Wittgenstein clearly brings mathematics into sociological focus: aren't proofs, he asks, parts of institutions; and aren't mathematical propositions anthropological propositions? More specifically, he contends that mathematics is *normative* in the sense of "norm", not in the sense of "ideal".[7] He links "inexorability" to socialization and social action; the meaning of a formula resides in "the way we always use it, the way we were taught to use it". And the inexorability with which two follows one and three follows two must be understood in terms of the series of cardinal numbers: " . . . in a different series something different follows. And isn't *this* series just *defined* by this sequence"? [8]

Mathematics is a *language game*. Wittgenstein stresses the importance of getting to the linguistic roots of what we say in mathematics. "Counting", for example, means "counting like *this*". We must not say of mathematics, any more than of language in general, that it is true; mathematics is *usable*, and, more importantly, *used*. Reading Russell and Whitehead's *Principia Mathematica* is a "game" for Wittgenstein: [9]

> I say that in Russell one proposition follows from another if the one can be derived from the other according to the position of both in a proof and the appended signs – when we read the book. For reading the book is a game that has to be learnt.

Proof is conceived in terms of use and custom. The laws of inference compel us in the same sense as other laws of human society. Wittgenstein suggests that mathematical propositions are "prophesies" in the following sense:[10]

> . . . they predict what result members of a society who have learnt this technique will get in agreement with other members of the society. . .

Wittgenstein conceives mathematics as problematic; he invites us to bring into question the unquestionable and to "Take a *wider* look around".[11]

LAKATOS

Lakatos' contribution consists primarily in undermining traditional history and philosophy of mathematics, and providing case studies which reflect the influence of his teacher, Karl Popper.

Mathematics is fallible; it develops through conjectures, criticisms, corrections, the search for proofs and the search for counterexamples. "Real" mathematics, mathematics as a process of growth and discovery, is *informal mathematics*. Philosophers have erred in attending to *formalized mathematics*, a species "hardly to be found anywhere on earth or in heaven outside the texts and journals of symbolic logic".[12]

Lakatos argues that "false philosophies" have distorted the history of science.[13] His case study of Cauchy illustrates this problem, and an alternative Popperian approach.

The problem Lakatos addresses involves the relationship between Cauchy's work on the calculus and the later developments due to Weierstrass. Cauchy was credited by historians of science with placing the calculus on a rigorous foundation and exorcising "false gods" such as, and especially, "infinitesimals".[14] Cauchy's continued usage of the term "infinitesimal" was interpreted as "a manner of speech". The development from Cauchy to Weierstrass was conceived as cumulative and progressive. But Lakatos asks how it is that fourteen years after the discovery of Fourier series Cauchy could "prove" that the Cauchy integral existed for any continuous function? Was this a matter of oversight, carelessness, or "'unfortunate' technical mistakes"? The conflict between Cauchy's theorem and Fourier's counterexamples was dealt with in a confused manner by Abel and Dirichlet. The problem was solved by Seidel twenty-six years after Cauchy constructed his proof. Why did a full generation of "the best minds" fail to solve this problem? It is not enough to point out that problems often look simple once they have been solved.

For in this case, Cauchy continued to defend his theorem even after the publication of Seidel's paper.[15]

The solution to this problem is suggested by Robinson. He points out that in the history of the calculus from Leibniz to Weierstrass there are two rival continuum theories: the theory currently accepted is Weierstrassian; the older and second theory is Leibnizian.[16] Cauchy was a Leibnizian. His theorem is true and his informal proof correct in the Leibnizian framework.[17] Thus Lakatos concludes that the appraisals of F. Kline, Cajori, Boyer and others regarding Cauchy's initiation of the Weierstrassian revolution are "utterly false", a "1984-wise rewriting of history according to the latest party line". The "error" attributed to Cauchy only appeared in Seidel's "translation" of Cauchy's proof from a Leibnizian into a Weierstrassian framework. Furthermore, Lakatos contends that even if Cauchy's proof had been formulated in a Weierstrassian context, to say (as Pringsheim did) that "the erroneous supposition that pointwise convergence *eo ipso* means uniform pointwise convergence" would still be "a characteristic piece of justificationist reconstruction". This reflects the widely held notion that informal proofs are formal proofs with "gaps". This view, according to Lakatos, leaves no room for the "genuine evolution of conceptual frameworks . . . " Leibniz's theory fell not because it harbored inconsistencies but because Weierstrass' theory had greater explanatory power and growth potential.[18]

Lakatos draws the following conclusions from this case: (a) past theories should be evaluated in context and not in terms of continuity with modern theories; (b) the strategy of maximizing continuity by demarcating a *hard*, formed, uncontested, indubitable, eternal core and a *soft* metaphysical, controversial, and changing interpretation of the formalism should be discarded; and (c) while foundation studies cannot be expected to lead to "ultimate rigour", it should be noted that they have on more than one occasion "stimulated and suggested further growth".[19]

Lakatos' case studies add some substance to Wittgenstein's remarks, and reveal a variety of inlets to the internal dynamics of mathematics where sociologists might find fruitful problems. In the next section, I briefly review the place of mathematics in traditional sociology of knowledge.

MATHEMATICS AND THE SOCIOLOGY OF KNOWLEDGE

Some indication of the lack of attention given to mathematics by sociologists of knowledge may be gathered by noting that in the index to the 1970 reader, *The Sociology of Knowledge*, edited by Curtis and Petras, there is no entry for mathematics.[20] Among the leading sociologists of knowledge who do remark on mathematics there is a tendency to treat it as a body of eternal truths independent of social forces. Thus Stark writes that "even the most abstract thought of the most abstract musician still stems from the prime value of the age, which shows itself as much in mathematics as in music, in science as in art".[21] But Stark does not go on to follow Spengler here: "It is nonsense", he argues, "to speak of bourgeois mathematics"; "... formal propositions and physical observations are not reducible to social and cultural phenomena and hence are free from relativity; it is nonsense to speak of historical relativity in mathematics and mechanics".

Znaniecki seems to be in agreement with Stark when he writes: "... the development of pure mathematics is not subjected to that relativity which characterizes all inductive theoretic knowledge of empirical data". But insofar as *mathematical physics* is concerned, Znaniecki argues: "The line taken by modern mathematical physics is only one of many, and it may break up sooner or later into several new, now unexpected lines". Znaniecki views pure mathematics as something different from knowledge in the scholarly or "modern inductive" sense; "it has no reference to any object matter beyond itself".[22] Znaniecki argues that as a result of the struggles between fighters for truth on one side and discoverers, systematizers, and contributers on the other, the range of "evident rational certainty" has been more and more narrowed. Mathematical certainty, he contends has been salvaged but at the cost of eliminating its ontological content. Znaniecki thus adopts the logicist position that "mathematics is absolutely certain only because and in so far as it is an extension of logic".[23]

Mannheim, as I have pointed out, exempted mathematics from the tools and perspective of the sociology of knowledge. There is no "mathematics" entry in the index to his *Essays on the Sociology* of Knowledge. And in *Ideology and Utopia* he argues, virtually in passing, that in mathematics, geometry, and pure economics the content is so formal and abstract that it seems to be "completely detached from the thinking social individual".[24]

BOS AND MEHRTENS

Bos and Mehrtens adopt a structural functional perspective on mathematics. Their view of the relations between mathematics and society is a compromise between the Spenglerian and Mannheimian views, though they lean toward Mannheim. Their social perspective is primarily *internalist*; mathematicians are conceived as "social beings", and mathematical development as "a process of interaction between mathematicians". They also argue that "the concepts of 'mathematician' and 'scientific community' have to be differentiated according to the historical period". Mathematics develops in a social context, and is affected by and influences society: "The social organization of mathematical activity defines the extent to which a free, unimpeded development of mathematical ideas is possible; and it determines which outside stimulations can act upon mathematics".[25]

The importance of social factors in the *externalist* sense varies, being dependent on the degree to which processes in mathematical development are "global" or "local". That is, the more local the processes, the more internalist the explanation will be for any given mathematical development; the more global the processes, the more externalist the explanation. The rise of a renewed mathematics tradition in the Renaissance is a relatively global process, and the influence of social factors in this case is high. The influence of social factors is low in the case of Poncelet's introduction of imaginary elements in geometry, a relatively local process. In between these two processes, Bos and Mehrtens place Huygens' studies on curvature (based on his studies of the pendulum clock and motivated by problems of longitude), and the somewhat more global developments of descriptive geometry (linked to military architecture) and operations research (linked to modern warfare).

According to Bos and Mehrtens, *social factors* can be the origin of problems (e.g., commercial arithmetic in medieval Italy), and can affect *working conditions* (e.g., the influence of operations research, cybernetics, game theory, and programming on the theory and practice of organizations), *styles of thinking* (e.g., the mathematization of jurisprudence in the seventeenth and eighteenth centuries), and *issues and values* (e.g., the interest in the infinite against the background of theological and philosophical themes). *Mathematics* affects society *indirectly* through physics and technology (e.g., the development from quarternions to vector analysis to electrical theory to electric technology), and *directly*, as in the application of mathematics to problems of taxes and finances, and the use of geometrical decorations. Mathematics also affects society through teaching in terms of problems treated, the social function and status of educational institutions, and through professions and trades in which mathematics is important.[26]

WILDER

The mathematician Raymond Wilder, influenced by the anthropological theories of Leslie White, has studied "the nature of the cultural forces affecting the evolution of mathematics". The first of his two books on this topic focuses on elementary aspects of number and geometry because their evolution exhibits all the characteristics "found in the development of more advanced mathematics". Following White, Wilder organizes his analysis around the central concept of *symbol*. He argues that the symbol system of mathematics is a specialized subclass of symbol systems used in the everyday world. Except for the rudiments of counting, there is little that is inevitable about mathematics as we know it today; mathematics is created by people, and is just as much a function of time-bound cultural demands as their other adaptations.[27]

In addition to the concept of symbol, the concepts "environmental" and especially "cultural" and "hereditary" *stress* play crucial roles in Wilder's analysis. Environmental stress refers to forces outside mathematics proper that stimulate growth, change, and decay. Hereditary stress refers to cultural stress *within* mathematics. One effect of the first type of stress is primitive counting. The crisis in geometry over the discovery of incommensurables among the Pythagoreans is an example of the second type of stress. Wilder concludes his inquiry by summarizing his findings in ten "laws" or "principles" governing the evolution of mathematical concepts. He also formulates a number of generalizations in the course of his analysis. I have *labeled* these principles for ease of identification. They are stated here as hypotheses. Those that appear among the ten "laws" are numbered according to their original sequence:[28]

1. UTILITY HYPOTHESIS. At any given time, there will evolve only concepts that are so related to the existing mathematical culture as to increase its utility in meeting the demands of either its own hereditary stresses or the environmental stresses from the host culture (# 1).

2. CULTURAL NOURISHMENT HYPOTHESIS. Needs of the host culture, especially when accompanied by increased facilities that may be provided for the nourishment of the mathematical subculture, will result in the evolution of new conceptual devices to meet the needs (# 6).

3. FICTITIOUS CONCEPTS HYPOTHESIS. As the need for new concepts is stimulated by cultural and/or hereditary stress, such concepts will be *invented* irrespective of realist criticisms that they are "fictitious" or "metaphysical".

4. ADMISSIBILITY HYPOTHESIS. The admissibility and acceptance of a concept will be decided by its degree of fruitfulness. In particular, a concept will not be forever rejected because of its *origin* or on the grounds of metaphysical criteria such as "unreality" (#2).

5. VIABILITY HYPOTHESIS. The extent to which a concept continues to be mathematically important is dependent both on the symbolic mode in which it is expressed and on its relationship to other concepts.

(a) If one symbolic mode tends toward obscurity, or even outright rejection of the concept, then – assuming the usefulness of the concept – a more easily comprehensible form will evolve (SYMBOLIZATION SUB-HYPOTHESIS).

(b) If a group of concepts are so related as to make consolidation of them all within a more general concept feasible, then the latter will evolve (CONSOLIDATION SUB-HYPOTHESIS) (# 3).

6. GENERAL CONSOLIDATION HYPOTHESIS. When two or more features of separate mathematical systems seem to complement one another, or look like they can be combined to form a more powerful or efficient theory, mathematicians will set about constructing them.

7. SOLVABILITY HYPOTHESIS. If the advance of a mathematical theory will be promoted by the solution of a certain problem, the conceptual structure of the theory will evolve in a manner to permit the eventual solution of the problem. The circumstances are then likely to manifest the CLUSTER HYPOTHESIS (# 4).

8. CLUSTER HYPOTHESIS. The solution of an outstanding problem is most likely to be accomplished by several mathematicians working independently, rather than by a single investigator (# 4).

9. DIFFUSION HYPOTHESIS. The opportunities for diffusion, such as may be provided by a universally accepted symbolism, increased outlets for publication, and other means of communication, will have a direct effect on the rate at which new concepts evolve (# 5).

10. SELECTION HYPOTHESIS. A variety of terminologies will be invented for any given mathematical development, and a generally accepted terminology will be *gradually* selected.

11. INNOVATION HYPOTHESIS. If a novel method is introduced to solve a special problem, it will subsequently be generalized to other situations and a new method of proof.

12. FORMALIST HYPOTHESIS. The formal manipulation of symbols may force the introduction of new concepts.

13. INVARIANT SEQUENCE HYPOTHESIS FOR CONCEPT FORMATION. Place/Denotation – Operation – Concept.

14. CRISIS HYPOTHESIS. New concepts will usually depend on concepts that are only intuitively perceived at the time, but that will ultimately generate crises due to their inadequacies. A crisis such as may be produced by exposure of an inconsistency in the current conceptual structure, or the inadequacy thereof, will stimulate accelerated evolution of new conceptions (# 8, 9).

15. GENERATIVE HYPOTHESIS. Solution of an outstanding problem will create new problems (# 9).

16. STAGNATION AND DECAY HYPOTHESIS. A static cultural environment will eventually stifle the development of new mathematical concepts. A similar effect will result from an adverse political or general anti-scientific atmosphere (# 7).

17. PROGRESSIVE HYPOTHESIS. Mathematical evolution remains forever a continuously progressing affair, limited only by the contingencies expressed in prior principles (# 10).

Wilder's second book differs from his first one in two respects: (a) culturological concepts are discussed in more detail, and (b) the analysis is broadened to encompass topics in higher mathematics.[29] The three basic ideas guiding Wilder's analysis of mathematics as a cultural system are (a) culture is a "super-organic entity" which obeys its own evolutionary laws; (b) evolutionary changes are implemented by environmental (external) and hereditary (internal) stresses; and (c) "Evolution", a process of change, must be distinguished from "History", a chronological record. Wilder tends to ignore complex and controversial conceptual issues in anthropology and mathematics in the interest of providing a brief, non-technical account of mathematics as an evolving sub-culture. The result is a discussion of development and change in mathematics that is sociologically interesting in spite of and not on account of Wilder's culturological perspective. The major limitation of Wilder's account is that although it adopts a threatening posture on platonism, it ultimately comes to the aid of "pure" mathematics.

Wilder's analysis is a parade of historical vignettes and episodes selected to exhibit cultural patterns in mathematics. These "predictive clusterings" include multiples, genius clusters, cultural lag, and patterns of growth and decay. They are summarized in a set of twenty-three "laws", with due acknowledgement of Michael Crowe's critique of his earlier work and Crowe's own laws of mathematical development. Wilder does not consider laws of nature or culture immutable; they reflect observed regularities and can be used to make more or less successful predictions.[30]

There are numerous tensions in Wilder's argument. Relativistic and pragmatic notions of mathematical truth which suggest that things might be otherwise in mathematics are wedded with a superorganicism that leaves individuals and organizations helpless pawns of mathematical fate. His culturology is dominated by a *mediational* view of mathematics as an activity that is hindered or helped by the "host" culture; but occasionally it wanders toward a more radical *constructivist* view of mathematics as the embodiment of the social organization of mathematicians.

Wilder misses numerous opportunities to dwell on the implications of taken-for-granted aspects of the social organization and culture of contemporary mathematics. For example, he points out that it took much less time for category theory to achieve widespread recognition than it did for the group concept because mathematical culture had become more homogeneous. He does not remark on the obvious dangers homogeneity poses for innovation in mathematics. Wilder states that "specialization inevitably spawns consolidation", but ignores the sociological conditions necessary for that particular type of specialization to develop. In discussing the cultural aspects of the development of non-Euclidean geometries he underscores Zeitgeist and overlooks the function of social networks. He appreciates world view analysis, but separates the core of "pure" mathematics from metaphysics, symbolism, methodology, applications, and the "host" culture (Law 14 is: "Revolutions may occur in the metaphysics, symbolism and methodology of mathematics, but not in the core of mathematics"). Wilder is especially insensitive to the ways in which social structural changes (e.g., bureaucratization and professionalization) within the mathematical sub-culture can transform mathematics.

In the end, Wilder defends mathematics-as-it-is, with all its social trappings. The mathematical sub-culture, according to Wilder, "works" and does not need to be – indeed, cannot be (given its superorganic nature) – tampered with. Wilder takes us out of the platonist Garden of Eden, but leads us into the land of structural-functional sociology and anthropology of science.

CROWE

Michael Crowe argues that Wilder and Lakatos have broken ground for a new *historiography* of mathematics. In the spirit of Wilder's search for generalizations, Crowe formulates ten "laws" of conceptual change in mathematics.[30] Unlike Wilder, he does not draw on sociological and anthropological resources but rather seeks to summarize "the insights of the new historiography of mathematics". His laws, therefore, basically reflect an internalist perspective. A couple of Crowe's laws reiterate Wilder's laws; laws 3 and 8 in Crowe are the same as laws 3 and 8 in Wilder. Of the remaining eight laws, seven complement Wilder. I will number them consecutively with Wilder's so that they may be incorporated into a Wilder–Crowe set of hypotheses:

18. IRONY. New mathematical concepts frequently come forth not at the bidding but against the efforts, at times strenuous efforts, of the mathematicians who create them (an example is Saccheri's creation of non-Euclidean geometry while trying to prove that no geometry but Euclid's is possible).

19. LAW OF RESISTANCE. Many new mathematical concepts, even though *logically* acceptable, meet forceful resistance after their appearance and achieve acceptance only after an extended period of time.

20. LAW OF RIGORIZATION. The rigor that permeates the textbook presentations of many areas of mathematics was frequently a late acquisition in the historical development of these areas and was frequently forced upon, rather than actively sought by, the pioneers in those fields.

21. MULTILAYERS LAW. The "knowledge" possessed by mathematicians concerning mathematics at any point is multilayered. A "metaphysics" of mathematics, frequently invisible to the mathematicians yet expressed in his/her writings and teaching in ways more subtle than simple declarative sentences, had existed and can be uncovered in historical research or becomes apparent in mathematical controversy.

22. THE LAW OF AUTHORITY. The fame of the creator of a new mathematical concept has a powerful, almost a controlling role in the acceptance of that mathematical concept, at least if the new concept breaks with tradition.

23. The TRANS-CONTEXTUAL LAW. New mathematical creations frequently arise within, and depend in the mind of their creators upon, contexts far larger than the preserved content of these creations; yet these contexts, for all their original importance, may impede or even prohibit the acceptance of the creations until they are removed by the mathematical community.

24. THE MONSTER-BARRING LAW (LAKATOS). Mathematicians have always possessed a vast repertoire of techniques for dissolving or avoiding the problems produced by apparent logical contradictions and thereby preventing crises in mathematics.

Crowe's tenth law is "Revolutions never occur *in* mathematics". This idea is pervasive in the literature on the history, philosophy, and sociology of mathematics. It reflects the difficulty even students of the social production of mathematics have with the apparent "eternality" of mathematical creations. Euclid, according to Crowe, was not deposed by non-Euclidean geometries. The stress on the preposition "in" in the "tenth law", Crowe points out, is crucial:[31]

> ... for, as a number of earlier laws make clear, revolutions may occur in mathematical nomenclature, symbolism, metamathematics ..., methodology ..., and perhaps even in the historiography of mathematics.

In a utilitarian, technical sense, Crowe is correct about Euclid. However, it is not clear he is correct when mathematical concepts or systems are construed as world views. I will discuss this further in Chapter 13.

FANG AND TAKAYAMA

Fang and Takayama focus on mathematics as a "happening-but-once" event. In contrast to Spengler, they argue that the " 'necessary and universal' concept of *numbers* has never changed anywhere at any time although the ways and means of *numerations* have".[32] The "happening-but-once" event they refer to is the development of Greek mathematics, that is, *Euclidean* mathematics. Their general sociological, quasi-Marxist conjecture is:[33]

> Society did, does, and will shape science, for there cannot but exist interactions – a truism! – between the *structure* (the economic base: forces of production and the relations of production) and the *superstructure* (the State and all the legal, political, and ideological or philosophical forms). The critical question is, however: to which extent? and in which manner? This question is all the more important because the "science" must belong to a sort of *super-superstructure* – the more abstract and the exacter, the higher – as Marx and Engels themselves did indicate.

In particular, Fang and Takayama argue that the sociology of mathematics is guided by three basic questions:[34]

(1) What social factors facilitated the creation and diffusion of the "happening-but-once" event of the birth of mathematics?
(2) Once created, what is the social role of mathematics and mathematicians in different times and places?
(3) How and to what extent do Science and Society shape each other in the evolutionary process?

While Fang and Takayama have been influenced by Marxism, they offer a very crude version of orthodox historical materialism. Their perspective (in particular, their notion of mathematics as a "happening-but-once" event) is virtually a caricature of the general tendency among Marxists to implicitly or explicitly give science and mathematics a privileged status among knowledge systems (i.e., more autonomy from the material forces that shape life and thought). Their interpretation of Greek mathematics as a "unique" event is consistent with "the Greek miracle" assumption which has influenced many historians of mathematics. In Chapters 14 and 15, I will show how a materialist analysis can help demystify "the Greek miracle".

FISHER

C. S. Fisher's study of invariant theory is based on the notion that "a theory is not a fixed object, but a social category which changes with the changing perspectives of mathematicians". Thus, the Theory of Invariants is constituted and evaluated differently in different groups of mathematicians, each with its own "mathematical Weltanschauung"; and classifications and evaluations evolve. Fisher cites T. S. Kuhn, G. H. Mead, L. Wittgenstein, A. Schutz, and A. L. Strauss among the scholars who have influenced his work.[35]

Fisher is concerned with accounting for the "death" or "disappearance" of the Theory of Invariants as a social category. Fang and Takayama have criticized Fisher on three points.[36] They note that the "death" of a theory may reflect "subsumption" or "integration". It is not clear, however, that Fisher's analysis precludes a "subsumption" or "integration" assumption. Fisher argues that the Theory of Invariants "died" as a "visible", "pure" tradition. He seems to leave room in his argument for the incorporation of the theory into newer developments, in particular in modern algebra. However, Fisher argues that while modern algebra could conceivably be thought of as a descendant of Invariant Theory, it is really "peripheral". Fang and Takayama's second criticism is more telling: Fisher ignores the renaissance in Invariant Theory from 1965 onwards; indeed, Fisher's work was completed on the eve of the renaissance. Their third criticism is that the *number* of people associated with a given theory is not a sound criterion for assessing the vitality of the theory. These criticisms notwithstanding, Fisher's study is important because it stresses the notion of theories as social categories, and relates the meaning of a theory to the Weltanschauung within which it is embedded. These ideas are not systematically developed, however; they are background to a study that is more social history than sociological analysis. The study falls short of the type of analysis Marxists have carried out, and at best only hints at Spengler's more radical world view analysis.

HAGSTROM

Warren Hagstrom is one of the few sociologists of science prior to the 1970s and excluding Marxists who has anything to say about mathematics. His study of the scientific community is based on a distinction between *formal* and *empirical* sciences. The formal sciences – mathematics, logic, and statistics – are distinguished from the empirical sciences by the fact that their theories are not, according to Hagstrom, subject to empirical proof. Furthermore,

Hagstrom argues that few of the problems mathematicians work on are given to them by "'nature' or practical exigencies". It is easy enough, however, to find students of mathematics – and perhaps more importantly *mathematicians* – who do not agree with Hagstrom's portrait of mathematics. Hagstrom himself cites opposing views. Von Neumann, for example, argues that the vitality of mathematics is dependent on a dialectic between abstract inbreeding and empirical tests. An algebraist interviewed by Hagstrom expresses discomfort with non-mathematicians who think "mathematics is cut and dried, simply a matter of finding things true or false". "In many ways", he says, "mathematics is more of an art than a science". But these are relatively mild disclaimers. A more thorough look at the world of working mathematicians (Hagstrom interviewed only about thirty mathematicians), including those who are by temperament and/or profession philosophically oriented, reveals that there is no general, stable consensus on matters of truth, falsity, proof, formalism, and purity. In addition to Von Neumann, H. Weyl, Russell, Gödel, and Quine are among those who have held that mathematics is an empirical, hypothesis-testing science. Mark Steiner flatly contradicts Hagstrom by claiming that "the notion of mathematics as a peculiarly 'formal' or 'deductive' science is mistaken . . . mathematics studies the natural numbers as zoology studies animals . . . ". Raymond Wilder, Morris Kline, Galois, Lakatos, E. H. Moore, A. N. Whitehead, and many others have at various times and in various ways challenged the purist, formalist viewpoint on the nature of mathematics. Even the relativity of *logic* has been defended by C. I. Lewis, H. Hahn, R. Carnap, and L. Rougier.[37]

Now none of what I have noted *shows* either that mathematics is or is not pure, formal, empirical, relative, or absolute. It clearly shows, however, that there is some variety of opinion about the nature of mathematics – within the mathematics community and among the community of scholars who study mathematics in philosophical, historical, sociological, and psychological terms; and there are controversies between and among these various communities. There is no need to deny the significance of consensus in mathematics to recognize that differences of opinion, conflicts, and controversies are crucial features of the mathematical landscape.

CONCLUSION

I have reviewed some of the major contributions to the development of the sociology of mathematics outside the Marxist tradition proper.[38] Some of these contributions bring mathematics into mainstream social history and structural functional sociology of science (e.g., Bos and Mehrtens). Others (e.g., Lakatos) are concerned primarily with giving a new account of mathematics from the inside, a new historiography of mathematics. Still others (e.g., Wittgenstein) point toward but fail to systematically exploit a sociological perspective on mathematics. The most vital developments in the sociology of mathematics have been, for the most part, the work of Marxists, and it is to their contributions that I turn next.

SECTION VI

THE LEGACY OF MARX

CHAPTER 9

DIALECTICS, MATERIALISM, AND MATHEMATICS

Karl Marx is one of the founders of what has come to be called "the sociology of knowledge and science". While Marx did not develop his ideas about knowledge and science systematically, it is easy to identify passages in his works which reveal his sociological perspective on science. He and Engels argued in general that even the simplest sensuous certainties are products of "social development, industry and commercial intercourse".[1] Elsewhere, Marx clearly sketched the idea of science as a social activity:[2]

> Even when I carry out *scientific work*, etc., an activity which I can seldom conduct in direct association with other men – I perform a *social*, because *human*, act. It is not only the material of my activity – like the language itself which the thinker uses – which is given to me as a social product. My *own* existence *is* a social activity.

The most significant remarks Marx formulated on the social roots of science occur in the 1844 manuscripts and the *Grundrisse.*[3] These remarks indicate that Marx conceived modern science as a bourgeois, alienated mode of inquiry. Thus, he associated the social transformation from capitalism to communism with a comparable transformation of modern science. Marx describes the transformation of science as the negation of science-as-it-is, and the emergence of "human science", dealienated, unitary (not "Unified"), wholistic, and global. The only science Marx examines in any detail is mathematics. But when he turned his attention to mathematics in his later years, it was not as a sociologist of numbers but as a student of the *conceptual* foundations of the calculus. He was especially interested in the idea of the differential.

In the *Mathematical Manuscripts*, Marx raised objections to the classical view of the derivative developed by Newton and Leibniz, D'Alembert, and Lagrange.[4] He labeled their approaches, respectively, "mystical", "rational", and "algebraic". Marx rejected the notion that the only significant thing in mathematics is the formal apparatus; his objective was to root mathematical operations in the "real world". Thus, he argued that ΔY and ΔX are "absolutely zero" when the derivative appears. Neither Newton nor Leibniz clearly stated this idea; and it was "accidental" (as opposed to foundational) in the works of D'Alembert and Lagrange. As Struik notes:[5]

The objection which Marx raised against the classical writers was that all four had the derivative already prepared before the process of differentiation really begins. Marx wanted a method which actually followed the process of variation of the variable and in this process itself defined the derivative as 0/0, in which case it can be endowed with a new symbol dy/dx. The derivative, he claimed, should be derived by a *process* of differentiation, not be produced from the beginning by the binomial theorem.

One of the reasons Marx appears to have anticipated later developments in the conceptual analysis of the calculus (already underway in his own time) is that he insisted on rooting the origin of the derivative in a real change and did not rely on infinitesimals. Marx also criticized the geometrical representation of the derivative and insisted on the operational nature of the differential. But his work on the calculus – while it might very well be appreciated by some mathematicians, and deals with recalcitrant mathematical issues – falls outside the trend toward arithmetization among turn of the century calculus students. And it is clear that infinitesimals were not yet banished in Cauchy's work, which is considered the basis for the modern rigorous treatment of the calculus, nor fairly treated in Georg Cantor's vehement but misguided arguments against infinitesimals, carried out while he was defending transfinite numbers. Indeed, infinitesimals are by no means outcasts in contemporary mathematics. Abraham Robinson, for example, has defended Non-standard Analysis as "a viable Calculus of Infinitesimals".[6] But the issue of how Marx's views on the calculus compare with those of mathematicians in his time and ours is tangential to the point that Marx was primarily attracted to the differential calculus because of its "dialectical qualities", and the possibility of giving a mathematical interpretation to "the negation of the negation". Not incidentally, he seemed intent on stripping "the veil of secrecy from science". Marx did not, however, give a sociological account of mathematical knowledge. Engels' "materialist" analysis of mathematics, however abbreviated and fragmentary, is more explicitly sociological.

Engels, like Marx, was intrigued by what he viewed as the dialectical nature of mathematics.[7] But in *Anti-Dühring*, he argues that mathematics, like the other sciences, arises "out of the *needs of men*", and in particular from the activities of measuring land and the "content of vessels", and "from the computation of time and mechanics".[8] As mathematics develops, according to Engels, it becomes more and more abstract, and the abstractions are more and more "divorced from the real world, and . . . set over against it as something independent, as laws coming from the outside, to which the world has to conform". But this development takes place in society; this accounts, ultimately, for the fact that "pure" mathematics can be *applied* to the "real world" – it is "borrowed" from the real world to begin with.

According to Engels, "the apparent derivation of mathematical magnitudes from each other" is accounted for by "rational interconnections" rather than "a priori origins". Pure mathematics deals with real "space forms and quantity relations". In an appendix to *Dialectics of Nature*, Engels argues that there are prototypes in nature for all the so-called "imaginary magnitudes" (such as differentials and infinities) which give the appearance of being "pure and exclusive" feats of the mind.[9] Thus, the mass of the earth seems infinitely large in comparison to the masses that are part of the everyday world; and the earth, as well as the solar system, appear infinitesimal in comparison to inter-stellar distances. Such differentials are the "real world" bases of the concept of infinity. It is not clear that Engels argument would be as nearly self-evident if he had to deal with twentieth century developments in mathematics and the philosophy of mathematics. Nonetheless, his attempt to critically evaluate the claims for "pure mathematics" and to outline a materialist reply is a notable achievement.

The next significant development in Marxist sociology of mathematics is Bukharin's comments on the history of mathematics. In the space of a few pages, and as an "aside", he draws on Moritz Cantor's history of mathematics to combat the common assumption that "mathematics, being a purely contemplative science, has nothing in common with practical life".[10] Bukharin notes that the development of mathematics in the ancient world was based on surveying, commercial arithmetic, vessel measurements, and calendar-making. He points out the close relationship between mathematics and religion; and the practical problems dealing with trade, household affairs, and everyday economics found in such texts as the *Calculation Book of Ahmes* (Egypt), the *Codex Arcerianus* (Roman Empire), the *Aryabhatta* (India), and various Chinese texts. This brief review leads Bukharin to the following conclusions:

(1) The content of science is given by the content of technology and economy.

(2) The development of science is determined in part by the *tools* of scientific knowledge.

(3) Social conditions can encourage or retard scientific progress.

(4) The economic structure of society determines the method of scientific thought.

(5) The class structure of society impresses its class stamp on the mode of thought and the form of material interest in mathematics.

The same "causal relations" are found in modern times, Bukharin argues, but in different and more complicated forms. Bukharin does not tackle the modern cases; nor does he discuss Greek mathematics. But it is these recalcitrant cases which are the most challenging. I will show later that in order to deal with the recalcitrant cases, it is necessary to go beyond the simplified (if not simplistic) view that equates the Marxist analysis of mathematics with identifying the prosaic origins of mathematical ideas and activities.

Spengler's chapter on number in *Der Untergang Des Abendlandes* is a major event in the history of the sociology of mathematics.[11] I mention it here because in a sense Spengler gives a more consistent materialist analysis of mathematics than the Marxists he was opposed to. His interpretation of "Cultures", "numbers", and "minds" suggests that numbers and number-meanings are imbedded in world views; there are as many number-systems as there are Cultures. Spengler argues, for example, that "Gothic cathedrals and Doric temples are *mathematics in stone*". He was aware of the problem of the limits of the "naturalistic" approach to number. He acknowledged the existence of certain characteristics which are apparently *independent* of Culture and century alongside "a quite particular necessity of form *conditioned* by Culture and century". He considered the problem of defining a frontier between these two types of characteristics, or even of demonstrating that such a frontier exists, insoluble. In any case, no Marxist theorist I know of has defended a comparably extreme thesis on the social nature of mathematics. Spengler's work is diametrically opposed to the Mannheimian tradition which holds that mathematics is immune to sociological analysis. Marxist studies of mathematics generally fall between these two extremes; at the same time, they manifest the tension that links the two polar positions. I will discuss Spengler's legacy in detail later in this book.

Unquestionably, the highlight of the 1931 International Congress of the History of Science and Technology in terms of the history of the sociology of science and mathematics was Boris Hessen's analysis of Newton's *Principia*. Hessen focused on the interrelations of economics, physics, and technology. He argued that the "abstract mathematical language" of the *Principia* does not reveal the connection between the problems Newton worked on and the technical demands they rest on, "the 'low' source of his inspiration". In order to expose this connection, it is necessary to recognize first that the geometrical arguments in the *Principia* do not represent the actual method of discovery (i.e., the calculus). While Hessen's work is a landmark exploration of the power of sociological analysis applied to science, Colman's paper on the crisis in the mathematical sciences is of more immediate interest.

Colman begins his paper with the following thesis:[12]

> The position of mathematics, as that of any science, is at bottom determined by the development and the position of the forces of production, of technology and economy. The latter affect mathematics both directly, by presenting it with new problems, creating its material basis and supplying its manpower, and indirectly through the prevailing outlook upon the world, the philosophy of the ruling class.

He argued that the crisis in the mathematics of his time was rooted in several basic "oppositions" or "contradictions": discreteness versus continuity; accident versus necessity; abstract versus concrete; historical versus logical; theory versus practice; and mathematics versus the foundations of mathematics. The resolution of the crisis would be effected, Colman wrote, through the "conscious, planned reconstruction [of mathematics] on the basis of materialist dialectics". To illustrate how he posed the contradictions and how he conceived the resolution, I quote his comments on discreteness and continuity:[13]

> We are thus presented with two extremes: continuity transformed into the absolute and the discrete made absolute, both seeking reciprocal unity, each giving a one-sided picture of material reality . . . the solution lies in creating a new manner of calculation which, proceeding from a uniform principle, unites in itself the positive sides both of analysis and discrete mathematics.

In commenting on the opposition between the historical and the logical, Colman considers the expression

$$\lim_{n \to \infty} (1 + 1/n)^n = 2.71828.$$

This expression was not, he argues, "artificially invented", that is, it was not the product of a logical development. It "arose from the *historical* requirements of overseas trade, from the need to create, for practical calculations, the most suitable logarithmic tables".[14] Thus, while Colman, like Marx, stresses the promise and implications of dialectical thought for understanding the *internal* dynamics of conceptual development in mathematics, he draws attention to specific links between mathematics and society more clearly than Engels. Another example is Colman's contention that analytic manifolds, built upon the concept of absolute discontinuity, "reflect with remarkable fidelity the extreme individualism of idealist philosophy".[15]

Marx and Engels established a basis for viewing mathematics as a social fact. Bukharin, Hessen, and Colman pioneered a Marxist sociology of mathematics in the 1920s and early 1930s. Scholars from Struik to Needham contributed programmatic and substantive studies in the 1940s and 1950s. But it was not until the 1970s that some continuity of research and theory in the sociology of mathematics was established.

DIALECTICS, MATERIALISM, AND MATHEMATICS

Dialectics. Efforts to apply or identify dialectical principles in mathematics and in the social relations of mathematics have not been as important or as successful as studies in the historical materialism of mathematics. Dialectical studies tend to be based on extremely general notions of contradiction and change, internalist (i.e., focused exclusively on self-contained mathematical symbols and operations), and procrustean.

The first important twentieth-century study of dialectics in mathematics is the one by Colman presented at the 1931 London conference. Colman, as I noted above, sets up certain polarities and conceives the contradictions and resolutions arising from these polarities as manifestations of dialectical principles. But he is so intent on revealing the dialectic at work that he fails to notice the extent to which he accepts arbitrary assumptions and concepts as self-evident characteristics of objective reality. This is generally true of students of dialectics and mathematics. The tendency to uncritically apply such terms as "negation" and "contradiction" in Marxist studies of mathematics is illustrated in an article by "Mike B.".[16] He argues that the process of solving the equation $x = 2\ 1/4 \times 30\ 1/10$ (where x is known *in form*) begins by taking logarithms. He describes this as "negating" the original equation. The resulting transformation gives: $\log x = 1/4 \times \log 2 + 1/10 \times \log 30$. This equation no longer states, even formally, what x equals. $\text{Log}\ x = 223$; the log table is used to "invert the transformed equation, returning it to its original form. It becomes $x = 1.67$. Making the original change and then removing it leads back to the original form. The contradiction is resolved". Mike B. argues that this is an example of "a negation of the negation". This seems to me a forced interpretation of dialectics. There is simply no self-evident reason for applying dialectical terminology to the problem solving strategy used in this case.

In another example, Mike B. describes the equation $3x + 4 = 19$ as one in which a contradiction is posed: "What is x"? He discusses this problem as follows:

> In elementary school, children memorize the mathematical law: "When equals are added to equals the results are equal", an expression of the dialectical interconnectedness of any equation. The movement which resolves the given contradiction begins by applying

this law and the equation $-4 = -4$ to obtain $3x + 4 + (-4) = 19 + (-4)$, or $3x = 15$. The oppositeness of addition and subtraction has been utilized to *negate* the 4, allowing it to move from one pole of the equation and penetrate the other, thereby changing both poles. A mutual penetration. We now use a version of the law with "added to" replaced with "multiplied by" and the equation $1/3 \times 3/1x = 1/3 \times 15$ or $x = 5$. This time the oppositeness of multiplication and division is used to negate the 3 and again mutual penetration takes place resulting in the solution of the equation.

Again, there is no self-evident reason for conceiving "What is x"? as a "contradiction", nor for using such phrases as "mutual penetration" to describe mathematical operations.

Struik's discussion of dialectical processes in mathematics is somewhat more satisfying than earlier or later efforts primarily because he conceives dialectical movement ("passing from") as a multidimensional phenomenon.[17] He fills out the framework for dialectical analysis constructed by Engels and Colman, but still leaves the impression that he is operating under a procrustean imperative. The problem seems to be that Marxists have been more intent on showing that dialectical principles really apply to the "objective world" (thus their generally misguided fascination with the calculus) than on constructing an adequate conceptual apparatus. A full discussion of this problem would take us far beyond the scope of this introductory essay. In my view, it would be useful to begin a reconstruction of dialectical analysis on the basis of Gurvitch's conceptions, which I outlined in Chapter 7. Gurvitch does not argue for the universal applicability of the dialectic. In his view, the dialectic applies to the natural sciences only to the extent and in such cases as they are affected by human factors. This conforms to Marcuse's interpretation.[18] David Bohm's dialectical view of *physical* reality discussed in Part I is in the tradition of those who like Engels conceive dialectics to be generally applicable to all aspects of reality.

Materialism. Chandler Davis, a materialist and a mathematician, has tried to resolve the apparent contradiction between materialism and mathematics. On the one hand, mathematicians think and talk about mathematical objects as mental constructs, or at least as objects that are different from everyday observable matter. The mathematician's main task is to make and consider "verifiably certain statements" about mathematical objects. All of this makes sense to idealists who claim that "things of the mind" have a primary existence, and platonists who claim that mental constructs exist independent of mind and matter. The problem for the materialist mathematician is this: as a materialist he or she claims that mental constructs do not exist "outside" of the real universe of things accessible to our senses; how is it possible to claim as a mathematician that certain knowledge about mathematical objects is possible through introspection?[19] Davis views this as a potentially fruitful "dialectical contradiction". Here is an outline of his program for a materialist mathematics.[20]

(1) Existence: a mathematical object exists if it can be talked about usefully – logically and syntactically – in mathematical reasoning.

(2) Definiteness: mathematical statements must be based on search procedures that are possible.

(3) Materialism: mathematics is a "human activity", a "living science" interrelated with other sciences; the products of mathematics are "human expressions".

(4) Recalcitrance or Facticity: the reason mathematical statements exhibit a greater degree of immunity to falsification than other types of statements is that they are *invariant* across "a comparatively wide range of alternative outcomes"; immunity to falsification is not a matter of mathematical statements being "*indifferent* to experimental results".

(5) Truth and Existence: there is a correspondence between mathematical existence and mathematical truth; a mathematical statement is true if it can be treated in mathematical reasoning the way a statement of fact is treated in reasoning about things in the observable world.

(6) Choice: a piece of mathematics should be judged (assuming it is formally valid) in terms of its potential for helping us make "factual statements about the objective world". (This, Davis admits, is a difficult criterion to apply in practice but it should be tried).

(7) Representation: becoming conscious of reality is not like imprinting a copy on a blank piece of paper; it involves reducing an overabundance of sensory inputs to "organized patterns of simple signals". Davis is not concerned with "crystallizing certified-true" mathematics, but rather with "doing" mathematics.

In a more recent paper, Davis traces the problem he addresses in his proposal for materialist mathematics to the changes in mathematics from the time of Georg Cantor on.[21] He thus associates the "peculiar platonism" of twentieth century mathematics with a mathematician who was a central figure in the professionalization of mathematics. Davis argues that twentieth century mathematicians are more "insulated" from other workers than any of their predecessors. According to Davis, this insulation (which in sociological terms is the result of specialization, professionalization, and bureaucratization) is the source of both platonist metaphysics and the rejection of applied mathematics.

Davis points out that the ruling classes tolerate and reward "pure" mathematics because in part mathematicians play a central role in maintaining the system of social stratification in advanced industrial societies. The "forbidding" nature of mathematics – its dryness, its difficulty, its insulation from human concerns – makes it a good selector of students with a high capacity for "self-discipline (sometimes called repression)". At the same time, the apparent objectivity of mathematics justifies the alleged impartiality of grading criteria. The tendency toward abstraction in the insulating sense is not, as Davis notes, peculiar to mathematics. It is associated with social structural changes dominated by specialization, professionalization, bureaucratization, and an increasing division of labor and a decrease in the interdependence of work activities, goals, and values. Davis thus links his program for materialism *in* mathematics to a materialist account *of* mathematics as a social phenomenon. In this way, he returns us to the mainstream in Marxist sociology of mathematics.

Dialectics and materialism *in* mathematics, and the dialectical materialism *of* mathematics and society, have received less attention than historical materialist analyses of mathematics and society. The fact that the former are less developed and less persuasive than the latter is not, therefore, surprising. More attention to the kinds of issues and problems I have remarked on in this chapter will be necessary before it is possible to evaluate the explanatory potential of dialectical materialism of mathematics and society, and of dialectics and materialism *in* mathematics.

In the 1940s, Dirk Struik, a Marxist, a mathematician, and a sociologist and historian of mathematics wrote a number of seminal programmatic essays on the sociology of mathematics. His work is an important link between the formative period in Marxist studies of mathematics and the period of research which began slowly in the 1950s and became an important part of the renaissance in social studies of mathematics in the 1970s. The following Chapter outlines Struik's historical sociology of mathematics.

CHAPTER 10

HISTORICAL MATERIALISM AND MATHEMATICS

Dirk Struik has narrated the history of mathematics from a materialist perspective. He consistently places mathematical developments in their social and historical contexts. For example, he links the evolution of arithmetic into algebra to the cultivation of number science in schools of scribes, and the development of the position system in Sumer to administrative techniques.[1] The relative stagnation of mathematics in the period preceding the emergence of ancient Greek civilization is attributed to cultural stagnation in China, India, Babylonia, and Egypt. But this is not left as an empty generalization. Struik traces the roots of the Greek emergence to political and economic changes in the vicinity of the Mediterranean basin during the closing centuries of the second millenium BC. The details of these changes are well-known: the bronze age gave way to the iron age; there were migrations and wars; the technology and social organization of war changed; production tools became less expensive; the social surplus increased; coined money and the alphabet were introduced. Struik underlines the importance of the growth in trade during this period. Constant struggle, in contrast to "Oriental staticism", stimulated the independence of merchants who recognized neither absolute monarchs nor static, all-powerful deities. These merchants enjoyed a certain amount of leisure (due to wealth and slave labor), and they could "philosophize". There were no established religions in the coastal towns, and this led many inhabitants to adopt mysticism; but it also fostered the growth of rationalism and science. Like some other students of the history and sociology of science, Struik stresses the significance of "democratic elements" as factors that promote critical thinking. Struik identifies the "sophists" as the first group of critical thinkers to approach mathematical problems with an orientation to understanding rather than utility. The sophists achieved this, according to Struik, because in an era of Athenian democracy they were "less hampered by tradition than any previous group of learned persons". The sophists emphasized the reality of change; the *aristocratic* Pythagoreans, by contrast, emphasized unchangeable elements in nature and society.

P. Tannery refers to problems surrounding the irrational and Zeno's paradoxes as a "logical scandal" in Greek mathematics. This crisis, Struik notes, is associated with the later part of the Peloponnesian war, and the fall of Athens in 404 BC.[2] There is thus a connection, he points out, between the crisis in mathematics and a social crisis that brought an end to slave-owning democracy and witnessed the emergence of a period of "aristocratic supremacy". In the early part of this new period, the intellectual atmosphere stimulated discussions of the foundations of mathematics and speculative cosmogony.

Struik correlates peace and the diffusion of knowledge in his discussion of the coexistence of the *pax Romana* and the *pax Sinesis* (under the Han). The differences between the lowly state of science in the Western Roman Empire, and the flourishing science of the Eastern empire are linked to extensive agriculture and wholesale slavery in the West and intensive agriculture in the East. Alexandria remained the center of ancient mathematics during the stable period of the Roman empire. But *compilation* and *commentary* eventually took precedence over original work.[3]

In Medieval Europe, feudalism and ecclesiasticism obstructed appreciation for mathematics (even in the monasteries) under northern and Germanic influence. The most important date during the early middle ages is probably 1085, when Toledo was taken from the Moors by Christians and students came into contact with treatises on science in Arabic.[4]

Algebra and computational arithmetic were stimulated during the early seventeenth century by the *Rechenhaftigkeit* of the mercantile bourgeoisie and (later on) surveying and navigation demands made by the leaders of the new national states; the productive use and further perfection of the machine was a related stimulus to mathematical development. There was, to a certain extent, a reaction against scholasticism; so-called Archimedian rigor was willingly abandoned, apparently without knowledge of the letter to Eratosthenes in which Archimedes used non-rigorous methods and "atomic" assumptions heuristically.[5]

Struik notes that mathematicians were quite active in the period prior to the Scientific Revolution. Cavalieri's work, for example, stimulated Fermat, Hudde, Barrow, and others to consider problems involving infinitesimals. In this "period of anticipation", discussion circles and constant correspondence were the means of communication and exchange. Certain persons, such as Mersenne, were centers of "scientific interchange". The academies crystallized out of the discussion circles, and in opposition to the universities which, aside from such exceptions as Leiden, were bastions of scholasticism.[6]

The seventeenth century was followed by a period of "wildly enthusiastic experimentation". Scientific activity, centered in the academies, was influenced by military problems, and problems in manufacturing: a great deal of

Euler's work, for example, focused on military questions and applications; and the Bernoullis were a merchant family. Euler was also among the many mathematicians who, under the influence of the lotteries and insurance companies organized during this period, became interested in the theory of probabilities.[7]

There are several points Struik makes in discussing this period that should be especially noted: (a) in regard to Berkeley's critique of the conceptual foundations of the calculus, he notes that this was not "the only case in which a critical difficulty in a science has been used to strengthen an idealist philosophy"; (b) the universal theory of gravitation was used as an anti-feudalism weapon by Enlightenment philosophers; (c) Maupertuis and others discussed the possibility of demonstrating the existence of God algebraically; and (d) the theory of probability was, in the Enlightenment spirit, applied to human concerns – probabilité des jugements – by Condorcet and later Laplace and Poisson. Struik identifies "deep-lying social reasons why English mathematicians refused to be emancipated from Newtonian fluxional methods". Victories in war and trade (England was engaged in constant commercial wars with France) and continental admiration of their political systems gave the English a "superiority complex". Mathematicians in late Alexandrian antiquity exhibited similar characteristics in an analogous setting. Toward the end of the century, some mathematicians felt that mathematics had reached closure. This pessimism was undermined by the French Revolution, and "the flowering of natural science" in France as well as (not surprisingly from a sociological perspective) on "the periphery of the political and economic centers", notably at Göttingen, where Gauss worked.[8]

In the nineteenth century, the French Revolution and the Napoleonic period were extremely favorable to mathematical growth; they:

(1) opened the way for the Industrial Revolution;
(2) stimulated the cultivation of the physical sciences;
(3) created new social classes;
(4) democratized academia;
(5) generated criticism of antiquated modes of thought;
(6) led to reform and rejuvenation in the schools and universities.[9]

Struik is equivocal on the issue of internal versus external factors in mathematical growth in the nineteenth century. He stresses the role of specialization in emancipating scientists from the demands of economic life and warfare ("science for its own sake"); and he notes that mathematics remained relatively sterile at the heart of the Industrial Revolution in England while it made great progress in France and later in Germany. But he writes in regard to "the most important contribution of this period", Gauss' surface theory in the *Disquisitiones generales circa superficies curvas* of 1827:[10]

> Here again practical considerations, now in the field of higher geodesy, were intimately connected with subtle theoretical analysis.

Of course, according to Struik, Gauss was "a genius" who broke out of the constraints of "specialization".

The work of "rigorization" carried out by Cauchy, Gauss, Abel, Bolzano and others parallels Eudoxus' rigorization of ancient geometry. There were also controversies during this period which invite study: Cantor was branded a "heretic" by Kronecker for his stand on the notion of infinity; and a "cult" emerged in support of the controversial "quarternions". By 1870, specialization had gotten out of hand. Reactions against it continue even today. The "enormous and unwieldly structure" of mathematics as of 1870 has yielded to syntheses which Struik believes constitute the most important contributions in mathematics over the last one hundred years.[11] The second part of the 19th century was marked by the great French comprehensive texts on analysis and its applications – the *Cours d'analyses*. Italian mathematics was reborn in the Risorgimento, and important contributions were made by Brioschi, Cremona, and Betti.

A SOCIOLOGY OF MATHEMATICS

Underlying Struik's view of the history of mathematics is a general sociology of science based on the following ideas:[12]

(1) There are, and we can discover, general laws of social development.

(2) It is possible to give a consistently materialist account of the world as a whole, including society.

(3) The discovery of the general laws of social development is pursued under the guidance of the following postulate: human productive systems and the social relations of production are primary shapers of social development.

(4) The visibility of outstanding ("aggressive") individuals should not obscure the relevance of class activities in science and society.

(5) Ideas, institutions, organizations, and other "material ideational conditions" enter into the laws of social development.

The general program for the sociology of mathematics is the study of how forms of social organization influence the origins and growth of mathematical knowledge, and the role of mathematics in the social formation of a given time and place.[13] The basic premises of the historical materialist approach are that (a) social and economic factors have a decisive influence on "the general character of mathematical productivity (or lack of productivity) in a given historical setting", and (b) form and content cannot be separated; but content eventually directs "the progress of form, deciding its importance and survival". Struik's various generalizations on the social relations of mathematics can be summarized as working hypotheses and assumptions;

(1) Sociological factors are the most important guide to understanding changes in mathematical knowledge during periods of "profound social upheaval, of transition, or in periods of widely varied social and cultural background . . . "; or, in a related formulation: "Mathematical ideas seem to develop on the edge of social struggles".[14]

(2) The separation of form and content, theory and practice, token world and objective world is a recurring phenomenon in the history of mathematics and society (e.g., during the disintegration of the western Roman empire; following the rule of Harsha in eighth century India; and during "the enfeeblement of imperialism"). This separation is "part and parcel of a general academic movement which finds in concessions to idealism the solution not only to its scientific and philosophic, but also to some of its social worries".[15]

(3) The following stages characterize the development or evolution of mathematics: (a) the empirical-practical stage (neolithic period; early civilizations of Egypt, India, Babylonia, and China); (b) the abstraction-via-generalization stage (already evident in the neolithic period); (c) further abstraction and the development of algorisms, theorems, and theories (for example, among the schools of scribes in ancient oriental religious and administrative centers such as Memphis, and Babylon); (d) the development of the logical structure of mathematics (beginning in Ionia and Greece; the test of truth becomes the absence of contradictions – in geometry, in arithmetic, and later in logic; this process reaches its culmination in the stress on consistency and completeness in modern mathematics); (e) the development of modern mathematics under the influence of capitalist growth; and (f) the emergence of mathematics as a specialized activity in the wake of the French Revolution.[16]

(4) "The logic of mathematics and especially of mathematics in its growth, is a dialectical logic"; passing from (a) the specific to the general (e.g., from the theory of equations of given degrees to Galois' theory); (b) the discrete to the continuous, the finite to the infinite (e.g., from sums to integrals); (c) quantity to quality (e.g., metrical to projective geometry to topology); (d) "real" to "ideal" domains (e.g., from arithmetic to the algebra of real numbers, complex and hypercomplex number systems); this is known

as the principle of the permanence of formal laws – two related principles are formulated: (e) the permanence of mathematical relations, and (f) the preservation of mathematical language.[17]

Struik resists sociologism and relativism. He argues that sociological, logical, artistic, and personal factors *all* play roles in mathematics. But like Marx, Struik constantly reminds us that mathematicians are always "social beings" even when they are worrying about such things as straight lines on hypercurves in seven dimensional space.[18] Struik also disassociates himself from relativism to the extent that he expresses a conviction that there is an objective, external world:[19]

> Our conviction of the eternal validity of Pythagoras' theorems, of the fact that $2 \times 2 = 4$, is not based on some *a priori* conception, nor can it be shaken by any clever mathematician who in a big book with formulas concludes that these theorems are mere convention. Our conviction is based on the fact that the theorems correspond to properties of the real world outside our consciousness which can be tested, and are accessible for testing to all persons from their earliest youth.

Very few mathematical discoveries in modern times are the result of "pure experimentation"; they are, however, developments within the framework of mathematics itself, not products of "pure" mental activity. Struik also argues that mathematics becomes a science when it goes beyond the empirical; that mathematical results are generally reached inductively; and that mathematicians, artists, and mystics share a common experience of harmony with the universe which expresses the agreements of thoughts with the real world.

There is a tension in Struik's work common to his Marxist predecessors between a conviction that "good" mathematics tells us about the "real world", and the idea that mathematics reflects particular social formations. His claim "that mathematics is possible, that it is applicable to nature, and that it has a social function are all interwoven" does not so much resolve the contradiction as it reflects an intellectual position rooted in Struik's commitments to mathematics, the idea of an objective reality, and materialism. The issue is this: how is it that particular societies can generate universal truths; or, how is it that modern mathematics is simultaneously "bourgeois" *and* in harmony with "objective reality"? Spengler seems to avoid this problem, as I will show later, but pays what some scholars believe is an exorbitant price-radical relativism.

CHAPTER 11

CONTEMPORARY MARXIST SOCIOLOGY OF MATHEMATICS

The tension in earlier Marxist analyses of mathematics between the idea that mathematics is a social product and the idea that it is in some sense independent of social forces continues to affect contemporary Marxist studies of the social roots of mathematics. The Bulgarian mathematician Iliev's paper on "Mathematics in Modern Society" in a special issue of the Polish quarterly, *Problems of the Science of Science*, illustrates this tension weighted in favor of the independence assumption.[1] Iliev emphasizes the internal dynamics of mathematics. This is treated quite technically and in some detail. The societal dimensions of mathematical development are mentioned, but only in passing. The following statements illustrate Iliev's orientation: (a) "Mathematics has served the needs of all the social and economic formations"; (b) "Mathematics has always been at the basis of technical progress"; (c) "Three important phenomena in mathematics are reflected in the development of all sciences and all conscious human activities of modern society: the construction and realization of mathematical structures, the mathematical foundations of computer technology and cybernetics, and simulation"; (d) The latter phenomena "and the technical problems they involve have their origin in the development of other sciences and society itself"; and (e) " . . . a change in the object of scientific research leads to a change in the internal laws of the development of science. Since contemporary society possesses means of directing science towards corresponding objects, we get the basic conclusion: modern society can influence the changes in the internal laws of the development of science. In particular, this is true for the mathematics also". While these statements imply reciprocity between mathematics and other aspects of society, on balance Iliev tends to glorify mathematics (it is, and is at the root of, "the most profound and brilliant achievements of human genius"), and argues in the end for the mathematization of society.

Luke Hodgkin's work is more sophisticated than Iliev's. He notes, to begin with, that even Marxists have trouble accepting the idea of social determinants in the history of mathematics.[2] However, he does not clearly distinguish between a Marxist analysis of mathematics and the type of demystification and demythification carried out by L. Wittgenstein and Imre Lakatos.

Hodgkin argues that scientific work, like all work, is tied to an "ideology of practice", an image of work, its significance, interesting questions, and fruitful approaches to problems. This ideology is socially determined, and in turn determines choices and orientations in scientific practice, and affects the *kind* of knowledge produced. The ideology of scientific practice arises in and is transmitted through institutions which determine choices, change ideologies, and determine what counts as scientific knowledge for the individual. Institutions mediate between "base" and "super-structure":

> It is through their influence that science retains and reproduces its character as *bourgeois* science – as knowledge adapted to a bourgeois society, whose forms of learning or work, or organization, are rooted in the division of labour and so . . . imitate the corresponding forms in society as a whole and are correspondingly oppressive.

The ideology of contemporary science, according to Hodgkin, is a bourgeois ideology. But at this point his analysis begins to manifest the tension between Marxist expectations about social roots and the recalcitrant "purity" and "autonomy" of science. The ideology of practice is bourgeois, Hodgkin argues, not primarily because concepts and theories in science reflect bourgeois interests and values, but because one of the functions of science "is to depict the activity of doing-science-within-capitalism as an acceptable choice for the individual". He then notes that "capitalism seems to survive best by having a number of competing ideologies at its disposal – in politics, in philosophy, in art; and the same is true of ideologies of scientific practice". These ideas are illustrated in a comparison of two French calculus textbooks.[3]

Cartan's *Differential Calculus* (1967) stresses abstract, theoretical understanding of why certain problems have solutions, and how to prove that solutions exist. Dieudonne's *Infinitesimal Calculus* (1968; the title, Hodgkin points out, is "deliberately 'reactionary'") stresses solving simple, practical problems. Hodgkin refers to these two approaches as competing bourgeois ideologies. Competition, he argues, is an important feature of contemporary science. The concept of "ideology of practice" is called on to explain the competition but not the content of the two approaches. It is not clear whether Hodgkin believes that competition of this sort is solely bourgeois. Hodgkin does *not* describe the *content* of the competing approaches as bourgeois. He points out that there is a "certain 'rightness' " in particular results which is not readily affected by social factors. This is true, he says, for the area of the surface of a sphere. However, there are differences in the frameworks for proofs and the rules of measurement used by Archimedes on the one hand and Stevin on the other in solving the problem of the area of the surface of a sphere. For Archimedes the solution appears in a proof that the area of a sphere is equal to the area of a related figure; for Stevin the solution appears as the formula $4\pi r^2$. Such changes, Hodgkin contends, are "named" in the ideology of practice. But again ideology is used somewhat indiscriminately to label the fact that concepts of proof and measurement *change* in mathematics; it is not used to reveal the relations between those changes and changes in modes of production or social formations.[4]

A similar problem arises in Hodgkin's discussion of the institutional sources of bourgeois ideology in science. It follows from Hodgkin's conception of institutions as mediators between base and superstructure that different images of science and scientific practice are associated with different institutions; therefore, science in a socialist society would be socialist science.[5] He shows, for example, how military and industrial requirements during World War II stimulated the development of coherent procedures for solving mathematical problems. The lack of coherence in early twentieth-century mathematics ("old mathematics") was noted by Colman in 1931. An example of an "old mathematics" problem taken from an engineering textbook is: "Solve the following partial differential equations as though they were ordinary differential equations". Hodgkin argues that the polarity between "knowledge" and "control" characteristic of nineteenth-century mathematics and twentieth-century "old mathematics" manifested in the separation of the pure theorist's role from the engineer/designer's role has disappeared. The "new mathematics" is centered around the achievement of "finite programs of work". But the claim that "pure" and "applied" mathematics were "separated" in the nineteenth century is not defensible in this bald form. Furthermore, Hodgkin's analysis of the development of the "new mathematics" is as much a contribution *to* as it is a statement *on the limitations of* Marxist analysis. He argues that it is impossible to view the new mathematics as a "reflection" of the economic base:[6]

> (And if I've chosen an example from "applied" mathematics – something with an obvious link (to production), it wasn't so much to make the connexion with the "base" more obvious as to make clear that they weren't the whole answer.) Production has determined the existence of the new mathematics; and maintains it in separate existence, but its forms, its guarantees of knowledge – in short, its scientific practice – are related to production in more indirect and complex ways (mediations), through the structure of the science's institutions, its accumulated history, its ideology of practice.

This is all quite reasonable, insofar as it expresses an awareness of the complexity of the relationships between base and superstructure. It is not clear to what extent Hodgkin conceives "economic base" in narrowly economic terms rather than broader social production terms; an awareness of the complexity of "economic base" is, in any case, needed in order to strengthen the approach Hodgkin sketches. The problem is that he avoids coming to grips with the implications of Marxist analysis for "pure" mathematics. It seems obvious that if *applied* mathematics is not connected directly to the "base", then *pure* mathematics must be even more, if not entirely, independent of the base. What is needed to clarify this issue is a better conception of the relationships between "base" and immediate work environments, and how these relationships affect the content of mathematics. While there is less tension on this issue in Hodgkin's analysis than in Iliev's, the tension remains nonetheless.

David Dickson's work introduces a more fully developed Marxist sociology of mathematics than can be found in Iliev or Hodgkin in two respects: (a) Dickson addresses the problem of explaining the *content* of mathematics more directly; and (b) he organizes his explanations more systematically in terms of basic Marxist concepts. In his study of the "metaphysical shift" that gave primacy to mathematics in seventeenth century Western Europe, Dickson poses the following questions: (a) how did the reduction of the "essential qualities of an object to a mathematical form" coincide with "the ideological requirements of the society" within which the reduction occurred; and (b) how were these requirements "reflected back in the particular type of mathematics that this society helped bring into being"?[7]

Capitalism depends on the systematic organization of production "as a complete system". The homogeneity between the various parts of the system necessitated by the need for systematic organization "can be expressed, for example, by reducing all components of the labour process to their economic 'value' ". There is a similar need for homogeneity in science that is "guaranteed by the use of mathematics":

> For mathematics is concerned with the relationships between objects (or events or phenomena) that can be demonstrated by translating them into one or another abstract mathematical language (the simplest being arithmetic, algebra and geometry).

The mathematization *of* nature can be viewed as a strategy for arriving at universally valid statements *about* nature. Commodities are fetishized "in order to make their appearance compatible with the general ideological forms – including 'wage', 'price', 'profit', etc. – in which capitalism articulates its social relations". The various aspects of the labor process must be reduced to quantified form before they can be dealt with in the capitalist process. There is thus "a formal correspondence between the way in which science subordinates the material world to the relations between mathematical symbols, and the way that capital subordinates the labour process to the relations between 'fetishized' characteristics . . . ". Dickson argues that from at least the seventeenth century on these two processes reinforce and express one another.

Dickson also argues that there is a "direct analogy" between the interchange which the calculus permits between "process and product" and the interchange between process and product "required by capital for the full articulation of, and control over, the links between the labour process and the commodity". Thus, the emphasis on controlling the labor process in the seventeenth century corresponded to a mathematics of process; earlier, the merchants' requirements in calculating commodity transactions had been met by the development of algebra:

> . . . the emergence of algebra in Western thought was part of the increasing activities (and often the political dominance) of a merchant class whose political power was based on the control of commodity transactions.

A fuller appreciation of Dickson's remarks will be possible after I discuss the relationship between commodity production, exchange value, fetishism, and abstraction in Greek mathematics, and capitalism and the calculus below.

By far the most ambitious project in the Marxist analysis of science and society is Joseph Needham's Science and Civilization in China project, now being carried on in the East Asian History of Science Library in Cambridge, England. Needham's problem is: (a) Why did modern science emerge in Western Europe and not in China or elsewhere; (b) this is considered a problem because during the fifteen hundred years prior to the Scientific Revolution, Chinese science and technology were in advance of Western science and technology in many respects; (c) the basic reasons for China's failure to undergo a Scientific Revolution, and for the West's success, are to be sought – Needham argues – in environmental (including ecological) and sociocultural factors. Needham treats mathematics within the general framework of this broader project.

Chinese mathematics was less systematic and abstract than Greek mathematics; the Chinese, however, were quite strong in algebra. They did not participate, according to Needham, in "the flight from practice into the realms of pure intellect".[8]

Some of the basic characteristics of Chinese mathematics noted by various observers are "the absence of the idea of rigorous proof" (Mikami), and correlated with this (according to Fu Ssu-Nien and others) the lack of a formal logic; and the dominance of "associative thought". Cajori remarks on the "beautiful symmetry" of the Thien Yuan algebra, as well as its "extreme limitations"; Cajori attributes the failure of Sung algebra to advance after the thirteenth century to a poor notation. Decimal-place-value and the zero "blank" were already in use by the fourth century BC. But " . . . no symbolic way of writing formulae was ever spontaneously invented by Chinese mathematicians . . . ".

Chinese algebra was "rhetorical" and positional. Symbols in the mathematical sense entered rarely and late. It utilized "abstract monosyllabic technical ideograms indicating generalized quantities and operations".[9] The counting board assigned certain quantities to certain places, establishing "a permanent filing-system of mathematical patterns". This was consistent with the "organic thinking" characteristic of Chinese philosophy. Equations were always connected with, and remained connected with, concrete problems; "no general theory of equations developed":

> However, the predilection for thinking in terms of patterns finally evolved from the counting-board a positional notation so complete (as far as it went) that it rendered unnecessary most of our fundamental symbols. Unfortunately, though the achievement was magnificent, it led to a position from which no further advance was possible.

Having thus sketched the basic characteristics of Chinese mathematics, Needham turns to a consideration of "the social background of Chinese mathematics".

The significance of mathematics in Chinese history is given in its relation to the calendar. This leads to one thesis about the concreteness of Chinese mathematics. The emperor established the calendar, and acceptance of the calendar by the tributary states was a sign of loyalty. Rebellions, famines, and other problems were often blamed on the calendar; in such circumstances, mathematicians would be called in to reconstruct it:[10]

> It has been thought that this preoccupation fixed them irretrievably to concrete number, and prevented the consideration of abstract ideas.

Mathematicians in China were "practical men" (see below): there is no record of "literary" mathematicians such as those found in Indian society. Chinese mathematicians were not able to interact with people such as the great alchemist, Ko Hung; Needham remarks that " . . . doubtless such thought connections would have been impossible anywhere before the Renaissance". Finally, "there was no room for Laws of Nature, and hence, no fixed regularities to which it would be profitable to apply mathematics in the mundane sphere".

The two most prominent dynasties so far as mathematical achievement is concerned are the Han and the Sung. In the Han, mathematics was clearly tied to government bureaucracy, "and devoted to the problems which the ruling officials had to solve (or persuade others to solve), such as surveying and land mensuration". Printing stimulated Sung mathematics, and "a group of truly great mathematicians, themselves either commoners or subordinate officials, broke out into fields much wider than the traditional bureaucratic preoccupations". Intellectual curiosity could now be pursued; but this did not last. It was followed and stifled by the nationalist Ming reaction; and the deadening essay examination system – the Pa Ku – was introduced in 1487.[11]

Needham argues that while it is true that China did not produce a Galileo, a Vesalius, or a Descartes, it did produce counterparts to Agricola, Gesner, and Tartaglia. One of the things that happened in Europe that did not happen in China, Needham notes, is that social changes made it possible and respectable for "gentlemen and technicians" to associate. The Galilean innovation which announces the Scientific Revolution, according to Needham, "may best be described as the marriage of craft practice with scholarly theory".[12]

Among the medieval Europeans, the higher artisanate had the experimental part of the Galilean method, and the scholastics forshadowings of the speculative part. Hypothesis making and testing were not, however, well-developed. There are parallels between China and Europe in the thirteenth century; the European scientific movement involving Roger Bacon, Bradwardine, Peregrinus, etc. had its parallel in the activities of Chhin Chiu-Shao, Yang Hui, Duo Shou-Ching, etc. There are parallels between scholastic and neo-Confucian thought – but neither mode of thought made clear the method of hypothesis.[13]

In the end, Needham argues, it appears as if only a mercantile culture could fuse nature-knowledge and mathematics, and stimulate mathematical growth; the agrarian bureaucratic civilization of China could support or sustain interest in Nature, controlled experimentation, empirical induction, eclipse prediction, and calendar calculations, but nothing more in science. The Chinese had every reason to pursue mechanics (for ship-building), hydrostatics (for their vast canal system), ballistics (for guns), and the science of pumps (for mines). Needham argues that they did not pursue these things because little or no private profit could be gained in them in a society dominated by an imperial bureaucracy. But in the West, mercantile culture reduces all quality to quantity, affirms a mathematical reality behind everyday appearances, and proclaims a uniformity of space and time everywhere in the universe – and, writes Needham, " . . . was it not analogous to the merchant's standard of value? No goods or commodities, no jewels or monies there were, but such as could be computed and exchanged in number, quantity and measure".[14]

MARXISM AND "THE GREEK MIRACLE"

The rise of scientific thought in Greece has often been referred to as "the Greek miracle". Clagett is critical of this notion, but still somewhat predisposed to it:[15]

> Study of the antecedents in earlier Greek and Near Eastern cultures does something to lessen the miraculous element but leaves us with considerable admiration for the first two centuries of Greek science and philosophy from about 600 BC to about 400 BC.

Clagett is hesitant about "social causes"; he refers to them as "uncertain, although probable". This contrasts sharply with Farrington's perspective on Greek science.[16] He argues that science originates in "the various activities by which man keeps soul and body together", "arises in contacts with things", "is dependent on the evidence of the senses", and "develops in close correspondence with the stages of man's social progress". This is close to the thesis on the economic foundations of abstract thought in Greece defended by George Thomson and Alfred Sohn-Rethel. The "pure reason" of classical Greece reflects the social relations of a *money economy; commodity production,* which blooms in ancient Greece, gives rise to the conception of nature as an independent object world; the process of *exchange* is the source of the abstraction inherent in commodity production and therefore of the abstract intellectual categories which permeate commodity producing societies.[17]

Thomson and Sohn-Rethel argue that (a) the original source of abstraction is commodity exchange, (b) that the formal elements of conceptual thought are contained in this abstraction, and (c) that the ideal abstraction invented by the Greeks (at least on an expanded scale) is rooted in the real abstraction operative in exchange. The argument runs as follows:

(1) Commodity exchange excludes use; therefore, it is abstract.

(2) Exchange is based on the notion that the physical aspects of commodities remain unaltered during the period they are not in use, that is, during the process of exchange. Marx pointed out that exchange value appears as something "*purely* conceptual", an "imagined entity" possessing "no *reality* except in the commodities, in so far as a definite amount of labour-time is materialised in them".[18]

(3) Time and space become continuous and homogeneous in exchange; use-value properties are "accidental", exchange-value properties, materiality ("the immutable and non-descript nature of the commodity"), are "substantive":[19]

As to economists commodities are composites of exchange-value and use-value, thus to philosophers all objects of cognition are composites of substance and accidents (sense-qualities).

(4) Use-value is relegated to private thought; and the participants in exchange conceive the relationship of ownership in mutually exclusive terms.

(5) Exchange involves (a) a principle of pure mathematical reasoning ("The equivalence of values in the exchange calls for changes in the *mode of quantity* when goods of different dimensional measurements are involved"); (b) atomicity; and (c) functional causality:

The commodities describe a pure linear movement through abstract – that is, empty, continuous and homogeneous – space and time as abstract substances which thereby suffer no material change and which are capable of none by quantitative differentiation.

(6) Commodity exchange abstractness is linked to "the thinking mind" through *coinage* in a difficult and lengthy process that stretches from Thales to Aristotle.

It is obvious that the efforts of "the thinking mind" depended on a division of labor that divorced mental and physical activities, and a class division that gave some people the "leisure" to reflect on and elaborate theory. Such people were not, however, divorced from the production process; they carried out the systematization of thought necessary for economic progress, that is, the rationalization of thought and the development of algorithms for easy, ready, and regular applications of ideas to production. Thus the Greek abstractions were not arbitrary but direct outgrowths of a geometry based on mensuration and surveying. In a manner similar to that in which social activities gave rise to religion, exchange gave rise to the abstract *conception* of exchange. Marx, in Volume I of *Capital*, writes:[20]

. . . the existence of the things *qua* commodities, and the value-relation between products of labour which stamps them as commodities, have absolutely no connexion with their physical properties and with the material relations arising therefrom. There it is a definite social relation between men, that assumes, in their eyes, the fanastic form of a relation between things. In order, therefore, to find an analogy, we must have recourse to the mist-enveloped regions of the religious world. In that world the productions of the human brain appear as independent beings endowed with life, and entering into relations both with one another and the human race. So it is in the world of commodities with the products of men's hands. This I call the Fetishism which attaches itself to the products of labour, so soon as they are produced as commodities, and which is therefore inseparable from the production of commodities.

This suggests why number and god are intimately linked, now obscurely, now vividly, from the Pythagoreans (and earlier) to the scholars of the middle ages (and later) who contributed to the development of the calculus through their speculations on the continuum, infinity, and infinitesimals in their religious quests.

Thomson links the following developments in Greek mathematics and science to the rise of the merchant class *intermediate* between the nobility and the peasantry, and commodity production and exchange:[21]

(1) The Pythagorean notions of (a) number as the primary substance, and (b) the fusion of opposites in the mean.

(2) The development of theory, divorced from practice, among members of a ruling class divorced from physical labor.

(3) Heraclitus' concept of the self-regulating cycle of perpetual transformation of matter.

(4) Heraclitus' "universal equivalent", fire: "All things", said Heraclitus, "are exchanged for fire and fire for all things, as goods are exchanged for gold and gold for goods".

(5) Oedipus, "the new man" in commodity-producing society.

(6) The Hebrew's "Jehovah" and the Parmenidean "One", both abstract, pure being, stripped of everything concrete and material.

(7) Pure reason as a fetish concept (a "reflex" of the money form of value).

(8) Aristotle's First Mover (a "reflex" of the system of owning "homogeneous slave labour").

(9) The Platonic theory of Ideas, and Aristotelian logic:

As we pass from Thales to Anaximander and Anaximenes, from the Milesians to Pythagoras and Herakleitos and finally to Parmenides, we find the concept of matter becoming progressively less qualitative and concrete, until Parmenides confronts us with a pure abstraction, timeless and absolute.

The extension of private property, commodity production and exchange, and the complete metamorphosis of exchange-value into an end-in-itself is associated with an intensification in the fetishism of ideas to the point that the social roots of ideas are not only obscured but denied or, ultimately, rendered invisible, inaccessible. Like wealth, ideas come to be viewed as things outside and independent of the human realm, things that can be characterized in terms of "absolute objectivity".

To the extent that *abstraction* is conceived as a process which leads to the discovery of things that are outside the realm of human activity, and outside of human control (we are, on this view, naive discoverers of the laws of nature, and not inventors), to that extent it is equivalent to *alienation*. And if the positive transcendence of human self-estrangement can only be realized by means of a "return" (through a renewed, necessarily more complex and therefore qualitatively distinct, mode of action and consciousness) to the *social* mode of existence, so alienated mathematics can only be transcended by returning it (and therefore ourselves) to the social mode of existence.

The commodities hypothesis is supported by Marx's treatment of *commodity*. His conceptualizations and definitions illustrate a correspondence between the elements of commodity production and exchange, and the elements of elementary arithmetic. The following statements by Marx can be readily paired with statements on the nature and function of elementary Euclidean propositions:[22]

> . . . commodities in definite quantities are congruent, they take one another's place in the exchange process, are regarded as equivalent, and despite their motley appearances have a common denominator.
>
> As objectifications of social labour, all commodities are crystallizations of the same substance.
>
> A certain amount of a spinner's labour-time is materialised, say, in 100 lbs. of linen yarn. The same amount of labour-time is assumed to be represented in 100 yards of linen, the product of a weaver. Since these two products represent equal amounts of universal labour-time, and are therefore equivalents of *any* use-value which contains the same amount of labour-time, they are equal to each other [things equal to the same things are equal to each other].

Marx's discussion helps explain why metals, and gold and silver in particular, play a central role in the development of exchange, and therefore, in the development of mathematics; for a given metal, there are no significant (from the exchange perspective) qualitative differences in internal composition or external appearance, and thus all relations involving metal are reduced to *quantity*. For a commodity to function as a measure of value (universal equivalent), it must be capable of expressing "purely quantitative differences, thus presupposing identical, homogeneous quality"; it must be divisible "into any desired number of parts"; and it must be possible to recombine these parts "so that money of account can be represented in palpable form".[23]

Marx discusses in some detail the queerness of a commodity; he refers to it as "abounding in metaphysical subtlties and theological niceties", "a mysterious thing". The equalization of different kinds of labor can occur only as the result of abstracting from their inequalities, "reducing them to their common denominator".[24]

To paraphrase Marx, the Greeks could not live on pure, deductive reason, or on axiomatic mathematics, just as the people of the middle ages could not live on Catholicism, nor the ancient world on politics:[25]

> On the contrary, it is the mode in which they [society as a whole, or by implication some part of it] gained a livelihood [e.g., the Greek philosophers] that explains why here politics, and there Catholicism [and here logic] played the chief part.

Simmel was aware of these connections. In his *Philosophie des Geldes* (1900), he argues that money economy initiated the ideal of exact numerical calculation, and that the theoretical counterpart of money economy is the exact mathematical interpretation of the cosmos.[26]

A CRITIQUE OF SOHN-RETHEL

The correspondence between Marx and Euclid noted above is suggestive on the assumption that commodity exchange and the Euclidean ideas arose at about the same time (over a three hundred year period, from 600–300 BC). But such a correspondence (other considerations aside) cannot carry the full burden of Sohn-Rethel's thesis. It is far from clear, for example, why exchange and coinage should be a source of abstraction, Marx's remarks on exchange-value notwithstanding, anymore than the elaborate (and coinless) credit systems (e.g., commodity-based transferable credit), "managed" legal tender currencies without metallic standards, and "substituted currencies" found in some of the early civilizations, including Babylonia.

In Marx's terms, the fetishism attached to the products of labor under conditions of commodity exchange is reflected in the fetishism which attaches itself to the products of thought. Abstractions, fetishized under conditions of alienation and class struggle, become "pure" ideas. But this can be interpreted as a rationale for proposing *religion* (or its social source) as the original root of abstraction.

The fact is that abstraction has to be quite advanced if language is going to develop. It is well-advanced in the primarily inductive mathematical systems of the ancient Babylonians, Egyptians, and Chinese. The type of abstraction Sohn-Rethel considers uniquely Greek is a necessary attribute of earlier societies. Angell, for example, argues that people had to develop at least three ideas *before* exchange could be developed and institutionalized: (a) private property, (b) relative values in exchange, and (c) measure.[27] The conceptualization of relative values is already a high order abstraction. In addition, mathematical theory was more advanced in Egypt and Babylonia than is generally recognized; and Greek mathematics was not as abstract as is commonly thought. Finally, Sohn-Rethel fails to take adequate notice of the fact that the peculiar type of abstraction he identifies with classical Greece is correlated with the ascendancy of Platonist oligarchism.

There is evidence that the Greeks of the Pythagorean era had achieved a level of abstraction that was comparable to that associated with the classical period but that was not systematized until Euclid prepared Books I, II, VII, and IX of the *Elements*. And the tunnel run through the hill of Kastro on the Ionian coastal island of Samos by Eupalinus of Megara was based on mathematical knowledge that was first systematically recorded six hundred years later by Hero of Alexandria.[28] These and other examples suggest that the capacity for abstract theory had reached a very high level too early to be accounted for in terms of the exchange abstraction thesis. In addition, it is probable that the commodity market, even at the height of Greek commercial culture was, like that of European commercial culture prior to the eighteenth century, limited. Given the added fact that a great deal of commercial life in the post-Ionian period was carried on by slaves, it does not appear that exchange was salient enough in the minds of the philosophers to support Sohn-Rethel's thesis.

There is an alternative to Sohn-Rethel's view of the "mystery" of coinage, and evidence that the Greeks did not in fact treat coinage as a vehicle of the exchange abstraction. Magic and religion are connected with the origin of money as a measuring-value unit. Gorini's analysis of the coin as blazon and talisman appears initially to provide support for Sohn-Rethel's thesis; for coins embody faith and magic and participate *simultaneously* in the physical and supra-physical world.[29] Gorini, however, argues that the Greek coin was both a means for acquiring concrete goods, and a talismanic power that was considered *just as concrete*. Furthermore, it was apparently quite common, even in the late Greek era, to speak of a given coin being worth so many oxen, rather than of oxen being worth so many coins.[30] Exchange-value failed to prevail over use-value even though commodity exchange was being carried on. The mystery of coins so far as the Greeks were conscious of it appears to have had more to do with magico-religious attributes than with the exchange abstraction.[31] Marx himself supports this conclusion.

Marx notes that when, in the sixth century BC, Anacharsis was asked what the Greeks used money for, he replied simply, "For reckoning". A couple of centuries later, things had changed but not as much as Sohn-Rethel assumes. Aristotle, for all his understanding of "commodity", did not grasp the idea that attributing value to commodities "is merely a mode of expressing all labour as equal human labour, and consequently as labour of equal quality". Marx suggests that this was inevitable given that Aristotle's Greece was founded on "the inequality of men and of their labour-powers". Significantly, while Aristotle recognizes the distinction between use- and exchange-value, he does not appear to recognize the possibility of producing something solely for exchange. Indeed, Marx characterizes "the writers of classical antiquity" as holding "exclusively by quality and use-value"; he finds no references to exchange value or to "the cheapening of commodities".[32]

Despite his well-conceived critique of Kantian *a priorism*, Sohn-Rethel fails in the end to sustain a materialist argument. He virtually separates "imagination" from its material foundations, and treats the "market" so abstractly as to eliminate it as a material basis for consciousness.[33] Mannheim's ghost is challenged, but not vanquished. Sohn-Rethel uses the exchange-abstraction to *defend* the objectivity of "exact science"; insofar as science is based on "the mathematical and experimental method" it is, according to Sohn-Rethel, "one and one only". He contradicts his criticism of Marxists who believe that science is ruled by timeless standards by arguing that Greek abstractions are universal but timebound: "there is a timebound cause for timeless logic". To his credit, he does not rule out the possibility that "in the long run the logic and method of science will alter as a result of socialist developments".[34]

SECTION VII

THE LEGACY OF SPENGLER

CHAPTER 12

NUMBERS AND CULTURES

Marxists have been prominent in the historical development of the sociology of mathematics. They have not, however, been completely successful in sustaining a consistent materialist argument. The reason for this is that they have been influenced by the pervasive notion that the more abstract forms of mathematics approximate (if they do not fully represent) universal and eternal relationships in *the* "real world'. But Spengler forces us to question the limitations of Marxism because his analysis of mathematics appears to carry the Marxist sociology of mathematics to its extreme conclusion. It is Spengler, not the Marxists, who stands diametrically opposed to Mannheim.

In Sections I and II, I reviewed the major developments in Marxist sociology of mathematics. My major criticism of Marxist studies of mathematics is that they are characterized by a tension between materialist assumptions and idealist or transcendental views of mathematical knowledge and truth. The resolution of this tension is a precondition for understanding the nature and limits of the idea that mathematics is a social product. Some efforts in this direction have been made, notably by Spengler. In spite of his antipathies toward Marx and Marxism, his idealism, and his rightist politics, Spengler's writings on mathematics and science are a fuller realization of materialist analysis than can be found in the Marxist program. My objective in this chapter, then, is to indicate why (or perhaps *where*) I think Spengler is a materialist, and why, furthermore, I think he is a more consistent materialist than the Marxists who have studied mathematics.

SPENGLER ON NUMBERS AND CULTURE

Spengler's discussion of numbers occurs prominently in chapter two of the first volume of the *The Decline of the West*. He uses *number* "to exemplify the way in which a Soul seeks to actualize itself in the picture of its outer world – to show, that is, in how far Culture in the 'become' state can express or portray an idea of human existence".[1] *Number*, "the primary element on which all mathematics rests", is *specifically* chosen "because mathematics, accessible in its full depth only to the very few, holds a quite peculiar position amongst the creations of the mind". Mathematics is "peculiar" because it is simultaneously a "science" ("fuller" and "more comprehensive" than logic), a "true art", and a "metaphysics". There is no *a priori* reason to agree with Spengler that mathematics is unique in this three-fold way; most if not all students of the sociology or natural history of mathematics assume mathematics is in some way unique among modes of knowing. What is significant is that Spengler's analysis is a formidable attack on the privileged status of mathematics as an autonomous intellectual discipline. Before proceeding further, it will prove helpful to *briefly* review some of the basic terms Spengler uses in his analysis of history.

Spengler makes several axiomatic distinctions: (a) "becoming" and "the become" (roughly, "process" and "result", or "experience as lived" and "experience as learned"); (b) "alien" (the outer world of sensation) and "proper" (the inner life of feeling); (c) "soul" (the possible, the future, "the still to be accomplished") and "world" (the actual, the past, "the accomplished"); and (d) "Nature" ("the numerable") and "History" ("everything unrelated to mathematics"): "An actuality is Nature in so far as it assigns things-becoming their place as things-become, and History in so far as it orders things-become with reference to their becoming". "Life" is "the form in which the actualizing of the possible is accomplished". As soon as these ideas are introduced, they are introduced as part of "an immediate inward certainty", as basic or elemental facts of consciousness. Waking-consciousness is conceived "structurally" as a "tension of contraries". There are two important points that should be noted about these contraries: (a) they are each units or totalities (and they are together a totality), and (b) they are polarities which by virtue of being extremes establish that there is a potential for many types of "realities"; "purely" organic and mechanical world views are at opposite ends of a series of "infinitely numerous and exceedingly heterogeneous" ways of seeing the "outer world". Finally, it should be noted that Spengler uses Culture in a specific sense that is different from the anthropologist's usage. When Spengler says that a "primitive man" has no Culture, he means that a "real knowledge of history and nature" is lacking. Only when the ensemble of self, history, and nature becomes separated for the "waking-consciousness" can we speak of Culture. In much the same way, Marx distinguished all human activity up to the threshold of communism as "prehistory", and the activities in communist society as the beginning of "human history". This does not mean that there has not been any history, in the conventional sense, up to the present.

In order to appreciate Spengler's notion of number, it is important to understand that he conceives of a "fundamental connexion between *the become* (*the hard set*) and *Death*". He then argues:

> The real secret of all things-become, which are *ipso facto* things extended (spatially and materially), is embodied in mathematical number as contrasted with chronological number. Mathematical number contains in its very essence the notion of a *mechanical demarcation*, number being in that respect akin to *world*, which, in the very fact of its comprising and denoting, fences off world-impressions. In number, then, as the *sign of completed demarcation*, lies the *essence* of everything actual, which is cognized, is delimited, and has become all at once – as Pythagoras and certain others have been able to see with complete inward certitude by a mighty and truly religious intuition.

Here we come upon the intriguing conjecture that mathematics and religion and linked. Such a linkage is implied, of course, in Spengler's world view perspective.

Number, according to Spengler, is "the symbol of causal necessity". Number and the conception of God both contain "the ultimate meaning of the world-as-nature". The deep affinity between religion and mathematics is clearly evident in the Pythagoreans and the Platonists. But it is present too in Descartes, Pascal, Leibniz and all of the great mathematicians. The "man who feels the spirit of number living within him – realizes that through it he 'knows God' ". Number is akin to God; and it is related to myth insofar as it originated in the "naming process" associated with the will to "power over the world". There is a clear rationale for Spengler's conjecture on the relationship between mathematics and religion in the cases he cites as well as in such cases as the relationship between the medieval discourses on infinity stimulated by theological questions and the development of the calculus, the differences between early British and modern algebra, and the relationship between theology and mathematics in Cantor's theory of aggregates.

Cultures, according to Spengler, are incommensurable. Our present minds, he argues, are "differently constituted" than minds in earlier Cultures. Therefore, earlier mathematical events should not be viewed as stages in the development of "Mathematics". The two major Cultures Spengler identifies are Classical and Western. He also identifies two minor Cultures: Babylonian–Egyptian and Arabian–Islamic (Indian and Chinese Cultures are also recognized in his schema). Each major Culture experiences the same birth-death sequence in its number-world: (a) *conception* of a new number form, (b) *zenith* of systematic development, and (c) *inward completion and conclusion* of a figure-world. In Classical Culture, the sequence is: (a) the Pythagorean conception of number as "magnitude"; (b) the achievement of the zenith between 450 BC and 350 BC in the works of Plato, Archytas, and Eudoxus: and (c) the inward completion in the works of Euclid, Apollonius, and Archimedes between 300 BC and 250 BC. In Western Culture, the sequence is: (a) the conception of number as "relation" (Descartes, Pascal, Fermat, Newton, Leibniz) in the seventeenth century; (b) the zenith achieved by Euler, Lagrange, and Laplace (1750–1800); and (c) the inward completion achieved from 1800 onwards by Gauss, Cauchy, and Riemann.

Classical mathematics deals with number as magnitude, as the essence of what can be perceived through the senses, that is, the essence of visible, tangible units; it is confined to facts in the present that are near, and small, with a focus on the properties of individual entities and their boundary surfaces (stereometry, or solid geometry); in general it is confined to positive and whole numbers and proportion as the nexus of magnitude. *Western* mathematics liberates geometry from the visual and algebra from magnitudes; numbers are images of "pure thought" (or, "desensualized understanding"), and their abstract validity is self-contained; the focus is on whole classes of formal possibilities, groups of functions, etc., that is, on relations, and on function as the nexus of relations. Whereas Classical mathematics affirms appearances, Western mathematics denies them; thus the opposition between fear of the irrational in Classical mathematics, and the central role of the analysis of the infinite in Western mathematics. In Classical mathematics, the straight line is the measurable edge; in Western mathematics it is an infinite continuum of points: and indeed the core unit of Western mathematics is the "abstract space-element of the point", and the main theoretical objective is the interpretation of *space* (a "great and wholly religious symbol", according to Spengler). Whereas enlargements and reductions of scale and the constancy of constituents are characteristics of Classical mathematics, Western mathematics is based on group transformations and the variability of constituents. In Classical mathematics, the equality sign in $3^2 + 4^2 = 5^2$ establishes a rigid relationship between specific magnitudes, and signals that a problem is being worked out to a result; in Western mathematics, $X^n + Y^n = Z^n$ is Classical in appearance but is in reality a new kind of number, a picture and sign of a relation – the equality sign does not point to a result in the Classical sense (and because of this, Spengler argues that a new symbolism is needed in order to eliminate the vestigial and confusing parallels with Classical mathematics).

Spengler's characterization of Classical and Western polarizes differences and so underscores his incommensurability thesis. There *is*, I would agree, an incommensurability between the Greek concept of number and the concept(s) of number fashionable among professional mathematicians from the late nineteenth century onward. But when Spengler contrasts Classical and Western, the latter label refers already to Descartes, Pascal, Fermat, and Leibniz. Earlier, the linkage between Classical and Western was much stronger. Stevin, called "the Archimedes of the Low Countries", and Valerio, "the new Archimedes", as well as other Europeans in the fifteenth and sixteenth centuries who uncovered and/or were influenced by the Greek corpus, used Archimedian methods. Actually, Spengler *is* aware of the fact that there was a *transition* in Western mathematics. He remarks on Vieta's introduction of a new symbolism in 1591 as "an unmistakable, if unintended, protest against the classicizing tendency of Renaissance mathematics". And indeed he writes: "The history of Western knowledge is thus one of *progressive emancipation* [my emphasis] from Classical thought, an emancipation never willed but enforced in the depths of the unconscious". But there is a less mystical explanation for this emancipation than Spengler offers in his references to "Souls" and "the depths of the unconscious". In fact, from the seventeenth century onward mathematics, like other activities in European culture, became increasingly specialized. As a result, mathematicians increasingly worked with the materials created within their own community. Each new generation worked with the manufactured symbols of the previous generation. This was the mark of the professionalization of mathematics, and helps explain why Western mathematics – especially from the nineteenth century onwards – becomes more and more removed from everyday "empirical observation", a point that seems to have eluded Spengler.

Let me finally summarize Spengler's notion of number as world view in his own words:

> We find an Indian, an Arabian, a Classical, a Western type of mathematical thought and, corresponding with each, a type of number – each type fundamentally peculiar and unique, an expression of a specific world-feeling, a symbol having a specific validity which is even capable of scientific definition, a principle of ordering the Become which reflects the central essence of one and only one soul, viz., the soul of that particular Culture. Consequently, there are more mathematics than one.

Spengler's thesis that there are as many number-worlds as there are Cultures can be conceived of in two forms: the weak form states that mathematics is a cultural phenomenon, that is, mathematical ideas and activities will vary from culture to culture. Let us briefly review some of these variations.

Autochthonous Chinese mathematics between the third and thirteenth centuries was essentially a "mathematics of survival". Chinese treatsies on mathematics deal exclusively with practical problems. The *Chiu-chang suan-shu* (*Nine Chapters on the Mathematical Art*), in existence as early as the third century BC according to some scholars, was a practical handbook for architects, engineers, officials, and tradesmen. It contains 246 problems on constructing canals, dikes, and city walls; taxation; barter; public services, and so on. The chapter on land surveying provides rules for finding areas, and for adding, subtracting, multiplying, and dividing fractions. Problems on the distribution of properties among partners, and on taxing goods of varying quality lead to arithmetical and geometrical progressions, all solved using proportions. Chapter 7 on problems in "excess and deficiency" introduces what Europeans later called the rule of false position for solving problems of the type $ax + b = 0$. Simultaneous equations using positive and negative numbers are introduced; and problem 18 in Chapter 8 involves five unknowns and four equations, a herald of indeterminant equations. The "Pythagorean theorem" (which the Babylonians had, and which appears in the *Chou-pei Suan-ching*, dated about 300 BC) is applied in Chapter 9.

Li Yeh's *Ts'e-yuan hai-ching* (*Sea Mirror of the Circle Measurements*), completed in AD 148, includes numerical equations up to the 6th degree; the procedure for solving such equations is not given, suggesting it was well known. The 170 problems in this treatise involve inscribing a circle in, and circumscribing, a right triangle. Chapter 1 introduces basic ideas about the inscribed circle. The results outlined in this chapter are applied in subsequent chapters to practical problems.

The *Shu-shu chiu-chang* (*Mathematical Treatise in Nine Sections*) was written in 1247 by Ch'in Chiu-shao, described by George Sarton as one of the greatest mathematicians of all time. There are nine sections in this treatise on: indeterminate analysis; astronomical, calendrical, and meteorological calculations; land measurement; surveying; land tax and state service; money and grains; structural works; military matters; and barter and purchase.

Other treatises, such as the *Yang Hui suan-fa* (*Mathematical Arts of Yang Hui*, first printed in 1378), and Chu-Shih-chieh's *Ssu-yuan yu-chien* (*Precious Mirror of the Four Elements*, written in 1303 at the end of the "Golden Age" of Chinese mathematics) follow the inductive pattern. In general, Chinese mathematicians were employed as provincial civil servants, or in the imperial bureaucracy. Deductive reasoning was by no means beyond their abilities; but a systematic, deductive mathematics is generally associated with the more or less specialized pursuit of mathematics over several generations. Ancient Greece achieved conditions conducive to the systematic development of deductive reasoning and "pure", "abstract" mathematics. The Greeks experienced changes in the nature of human productive activity that did not occur to the same degree or on the same scale in China; these changes fostered levels of specialization and generational continuity that furthered the systematization of mathematics. Chinese mathematicians were not specialists to the same degree that some mathematicians in ancient Greece were, and certainly not to anything like the degree achieved by mathematicians in Western Europe from about 1600 onward. The degree of specialization and autonomy in the social activity of mathematicians may be related to the relative weight given to inductive versus deductive reasoning, the preference for algorithms over abstract theories, and even (in a more complex manner) the predominance of algebraic over geometric thinking.

In the case of the Greeks, social conditions fostered the development of an elitist geometry opposed to everyday household and commercial arithmetic. As the separation between hand and brain, between basic sustenance labor and elite philosophy, progressed from the Pythagoreans onward, the unique "purity" and "abstractness" of Greek mathematics progressed apace. But economic diversity and exchange never reached the levels they were to reach in Europe from the eighteenth century on; there were correlative limitations on specialization, professionalization, and the length of teacher-student chains for transmitting the results achieved by one generation on to the next; this sort of generational continuity places each succeeding generation in a world of specialized symbols that become the materials of work – work that is progressively removed from everyday reality and thus increasingly purified and abstract.

The horizontal density of mathematicians in Greece (measured in terms of the number of generations of mathematicians linked together) and the vertical density (the number of mathematicians working in a given period) were generally higher than in China. There is a notable tendency to lose track of past advances in Chinese mathematics. Horizontal density is low in China both in terms of the lack of an ancestral community such as Europe had in the Greek corpus (the resurrection of Apollonius, Archimedes, Aristarchus, Euclid, Eutocius, Hero, Pappus, Ptolemy, Serenus, and others from the fifteenth century on), and in terms of the lack of generational continuity through schools and teacher-student chains. There is not a word about Sung mathematicians in the *Yu-Hai* encyclopedia of 1267, and nothing on Sung and Yuan mathematicians in the *T'ien-chung chi* encyclopedia of 1569. More or less independent mathematicians appear for the first time in the Sung dynasty. They are mostly (with the notable exception of Shen Kua) wandering plebians or minor officials. These scholars broke away from bureaucratic preoccupations, but they were not teachers and did not found schools. Their works were never, apparently, used in the Office of Mathematics (Suan-kuan), founded as early as Sui and T'ang for training minor officials in practical mensuration; and their works appear to have been unknown in the Ming period. For these and other related organizational reasons, the Chinese did not systematically develop Euclidean-type proofs.

Puzzles, competition, and proofs were not significant features of the oriental mathematical landscape, relative to western mathematics. But there was some interest in proofs in the Islamic-Arabic world during the period AD 700–1000 (and later to some extent). This interest reflects an increase in the density of the mathematical community accompanied by some competitiveness (there is some evidence, for example, of rivalries: Al-Kindi versus the Banu Musa and Abu Ma Shar, for one; and the priority dispute over who discovered the sine theorem, Al-Khujandi, Abu-l-Wefa, or Abu-Nasr-ibn-Ali-ibn-Iriq). Scholars such as Thabit-ibn-Qrra developed an awareness of proofs as they translated and worked with the Greek texts. The lower level of interest in, development of, and originality in proofs among the Arabic-Islamic mathematicians in comparison to the Greek mathematicians reflects the fact that the Arabic-Islamic community was less dense than the Greek community, less competitive, and not as well-organized in terms of teacher-student chains and schools.

Hindu mathematics, especially in the period before the influx of Greek astronomy (ca. AD 400) placed a unique emphasis upon large numbers. The wisdom of Buddha is illustrated by the gigantic numbers he can count out (on the order of 8 times 23 series of 10^7), and his magnificence is illustrated by the huge number of Bodhisattvas and other celestial beings who gather to set the scenes for his various sutras. The Hindu cosmology includes a cyclical view of time which is based on great blocks of years called yugas. There are four types of yugas ranging from 432 000 to 1 728 000 years, all of which together make up one thousandth of a kalma, or 4 320 000 000 years. This emphasis upon immense, cosmological numbers is the basis of a distinctively Hindu view of the near-infinite stretches of being that surround the empirical world. This is mathematics for transcending experience, but not in the direction of rationalistic abstraction. Numbers are used to mystify and impress; they are symbols in a mathematical rhetoric designed to stimulate religious awe. The social roots of this distinctive use of numbers may be found in the particularly exalted status of Indian religious specialists relative to their publics. The concrete as opposed to abstract nature of Hindu large numbers may also have been suggested by a social reality: the great variety of ethnic groups making up Indian society, institutionalized in the ramifications of the caste system.

The fact that different cultures exhibit different mathematical patterns does not support Spengler's incommensurability thesis. The differences are compatible with the notion of a long-term development of "universal" or "world" mathematics, nourished by the "true" results of all civilizational efforts in mathematics; the "false" results are gradually eliminated and forgotten in the main stream of mathematical evolution. The strong form of Spengler's thesis – the world view thesis – goes beyond the assertion that mathematics in different cultures varies, and challenges the notion that all civilizational traditional in mathematics flow into the "universal" mathematics of modern Western civilization.

CHAPTER 13

MATHEMATICS AND WORLD VIEW

Spengler, more than any other student of mathematics as a sociocultural phenomenon, carries us to, if not over, the threshold of a sociology of mathematics that is free of transcendental and idealist assumptions. Bloor notes that the main barrier to a sociology of mathematics is "the idea that mathematics has a life of its own and a meaning of its own".[1] He is sympathetic to Spengler's viewpoint; but his own version of a naturalistic approach to number contains some contradictions. He believes, on the one hand, in an eternal, material world that is the source of "permanent" truth. On the other hand, he argues that belief in a material world "does not justify the conclusion that there is any final or privileged state of adaptation to it which constitutes absolute knowledge or truth".[2] This statement is at odds with what Bloor has christened "*the strong program* in the sociology of knowledge". The strong program is (a) causal (what conditions "bring about belief or states of knowledge"?); (b) "impartial with respect to truth and falsity, rationality, or irrationality, success or failure", (c) symmetrical: true *and* false beliefs are explained in terms of the same types of causes, and (d) reflexive: the explanatory patterns in the strong program apply to sociology itself.[3] Bloor's "methodological relativism" is unhinged, however, by the scientistic basis of the strong program: "only proceed as the other sciences proceed", Bloor asserts, "and all will be well".[4] Here Bloor parts company with Spengler. For the strong program relativizes everything except science; this biases the reflexivity provision of the strong program by making it resistant to criticisms of and radical changes in science. Bloor thus fails to realize the world view implications of Spengler's thesis. Spengler allows for the emergence of a new Culture and a new world view incommensurable with the Western Soul or Culture, and therefore for a new kind of science or a new mode of inquiry.

Spengler anticipates the meta-analysis of science as a world view developed in the works of Radnitzky, Hooker, Bohm, and Feyerabend discussed in Chapter 7. The idea that science, scientific ideas, and scientific theories can be construed as world views is implied in one of the various interpretations Kuhn gives to "paradigm".[5] This is as close as anyone whose work is widely known and influential in the sociology of science and knowledge has come to defending a world view perspective on science. The idea has taken root outside of the sociology of science and knowledge proper.

TOWARD A SPENGLERIAN SOCIOLOGY OF MATHEMATICS

Hofstadter distinguishes between the *syntactic* and *semantic* properties of an object:[6]

> . . . the syntactic ones reside unambiguously inside the object . . . whereas semantic properties depend on its relations with a potentially infinite class of other objects, and, therefore, are not completely localizable.

On the basis of this distinction, it might be argued that "2 × 2 = 4" has localizable properties because it symbolically maps an intractable reality. But it also has semantic properties which human beings "assemble" over time – the meaning of "2 × 2 = 4" is based on its "connections to cognitive structures", and changes as cognitive structures change. Since no symbol or statement is ever created independently of cognitive structures, objects – as Hofstadter essentially concedes – can *only* have semantic properties.

In Bohm's "rheomode" (see Appendix C), movement is primary, and is reflected in the primacy of the verb. In "mathematical rheomodation" the focus is on "the mathematical aspects of the rheomode, considered as an unrestricted totality". Bohm writes:[7]

> . . . it is important to be alert to the fact that the present form of mathematics is genera-ally strongly attached to the ordinary mode of using language, with its function of emphasizing division into separate things of a static nature. Thus, consider such key mathematical words as all, some, none, set, identity, equality, etc. These evidently do have strongly static and divisive implications that would not be consistent with rheomodation. It will be necessary then to develop mathematics in a new way, allowing us to think in a "flowing" mode, which nevertheless consistently assimilates that which is essential in current mathematical concepts.

The last sentence in the preceding quote suggests that Bohm views certain aspects of mathematics as at least relatively static ("essential") if not as universal, eternal truths. But notice that these essential aspects become *assimilated* into any mathematical innovation under rheomodation. Since the changes Bohm has in mind are by his own admission world view changes, the assimilation of current mathematical concepts implies fundamental changes in the *meanings* of the assimilated concepts. Let me give substance to these ideas by examining the general notion of mathematical truth as a world view variable.

During the course of the nineteenth-century, as Joan Richards has noted, the concept of mathematical truth changed from "an objective truth about the real world" to "an abstract truth based on completeness and consistency within a formal mathematical structure".[8] Historians of mathematics have searched for the roots of this change in nineteenth-century mathematics, and one of the cases they have examined in detail is the development of algebra in early nineteenth-century England. The problem raised by the English case is this: why was it that the English algebraists were able to clearly enunciate a concept of abstract algebra but not to actually construct such an algebra? The answer lies in the "context" of early nineteenth-century English algebraic work. "Context", as Richards points out, must be interpreted *broadly* as world view:[9]

> The view of mathematical truth which informed the English development of the foundations of algebra was an integral part of a perception of truth which permeated their view of the world.

Mathematicians such as George Peacock, Augustus De Morgan, William Whewell, and John Herschel were members of a group of intellectuals Susan Cannon refers to as "the Cambridge Network". These men, Richards writes:[10]

> ... shared a perception of human intellectual potential. Briefly, they believed in a single and absolute truth which was the goal of all human investigation in any area. Thus, for example, the truth about the nature of man to which one aspired in a biological investigation was essentially complementary to that truth which was sought in a religious context.

The development of the abstract notion of truth occurred in the context of a world view change. In the absence of a rigorous meta-analysis, this change can be characterized in terms of the fundamental sociocultural processes in which it was rooted; these include changes in social organization (e.g., specialization) and secularization. Let me briefly discuss these two processes and their impact on the mathematical community.

Early in the nineteenth century, important organizational changes began to emerge in the European intellectual community. The organizational structure of the scientific world was becoming pluralized.[11] Competition became increasingly institutionalized within a mathematical community that was undergoing specialization and professionalization. The controversy between Cantor and Kronecker carried on the tradition of interpersonal competition. In fact, while Kronecker was better placed than Cantor politically (he held a chair at Berlin, edited Crelle's Journal, and was active in the

European mathematical community as a member of the Paris Academy, the Royal Society, and other organizations outside Germany), Cantor exhibited the same sort of "robber baron" behavior in opposing work on infinitesimals that Kronecker exhibited in opposing Cantor's work on transfinite numbers. But their controversy culminated in an organizational transition.

By the time Kronecker died in 1891, the controversy between Cantor and Kronecker had already begun to shift from the interpersonal level to the collective level, and to a new and higher level of social organization than had been known or possible earlier. The situation in Germany was a reflection of and a major arena for social structural changes in mathematical activity. The autonomy of mathematics as a profession was reflected in and fostered by the founding of the Deutsche Mathematiker-Vereingung in 1891 with Cantor as its first president. Cantor was also involved in promoting the First International Congress of Mathematicians, held in Zürich in 1897.

Psychologically, perhaps, Cantor's efforts may have reflected a need to break away from the power of the Berlin "Brahmans", Kronecker and Weierstrass, and to get himself a wider hearing, especially among younger mathematicians. Sociologically, these efforts were carried out in the context of the increasing density of the mathematical community, and the mathematicians' struggle for professional autonomy.

Kronecker, in his refusal to accept an actual infinity, his insistence on defining mathematical concepts in a finite number of steps, and his opposition to irrational, transcendental, and transfinite numbers, can be considered a precursor of the intuitionists. Cantor, by constructing his mathematics on the generalizations and abstractions of his predecessors and adhering only to the norm of consistency, expressed what was to become the formalist program (though without the degree of axiomatization and rigor demanded by Frege and others). Frege seems to have sensed what was happening sociologically. He argued that once the logical foundations of Cantor's world were established, the significance of transfinite arithmetic would become apparent. In 1892, Frege predicted a major conflict between two irreconcilable camps which would affect all of mathematics; those who affirmed and those who denied the existence and logical soundness of the infinite. The crucial role of infinity in this conflict is a clue to the significance of secularization in the mathematical changes I have been discussing, since historically studies of infinity and God have been closely linked.

Recall that the English algebraists had an integrated view of the world within which mathematical and theological truth (along with other truths) were "one". This "unitary view of truth" was broken down by such processes as specialization. And in the larger context of secularization, mathematical and theological truth became separated. The conflict between Cantor and Kronecker, in fact, appears to be part of a crisis within mathematics generated (at least in part) by the separation of mathematical and theological truth. The so-called "foundations" crisis which has played a central role in the development of twentieth-century mathematics, may be essentially a crisis based on the need to re-integrate mathematics into a view of the world in which the theological component is eliminated or radically altered.[12]

The notion of mathematical truth as world view illustrates the *potential* in world view analysis for resolving the problem of the limits of the sociology of mathematics. It does not, however, clearly and unequivocally transfer meaning to a statement such as "2 X 2 = 4 is imbedded in and reflects a world view". The best I can do at present is suggest what such a statement might mean. For example, "2 X 2 = 4" is a condensation of thousands of years of human experience *within* a particular community of consensus. This community is virtually universal, but earth-bound. There are situations and communities of consensus (especially in higher mathematics) for which "2 X 2 = 4" in its conventional sense is not meaningful.[13] It cannot *mean* the same (in the semantic sense) to a pre-Socratic and to a modern mathematician familiar with modular algebras.

Given the statement "2 X 2 = 4", one would have to establish a whole set of conditions that are usually taken for granted (e.g., the existence of certain prerequisties for arithmetic manipulations such as "normal" cognitive functioning) before its meaning could be specified. The fact that most if not all humans and cultures meet those conditions *does not* make "2 X 2 = 4" a "pure" or "transcendental" truth. Nor does the fact that certain experiences and statements may *bridge* world views or that world views may overlap, contradict the Spenglerian viewpoint on number.

EXEMPLARS: BLOOR AND MACKENZIE

Earlier, I pointed out that Bloor's "strong program" is a constraint on the extent to which he is able to realize a Spenglerian sociology of mathematics. But his programmatic statements may not, after all, hinder his research. Indeed, his suggestions about classification, network theory, and grid-group theory are important contributions to the development of a materialist sociology of mathematics.

In 'Polyhedra and the Abominations of Leviticus', Bloor draws attention to a common theme in Lakatos' *Proofs and Refutations* and Mary Douglas'

Natural Symbols.[14] Lakatos' book is a description of the history of a mathematical dispute and a philosophy of mathematics. Douglas' work outlines an anthropological theory of pollution, ritual, dietary restrictions, and religious cosmologies. Both books, Bloor claims, "deal with the way people respond to things that do not fit into the boxes and boundaries of accepted ways of thinking"; they are about anomalies that do not "fit" publicly-accepted schemes of classification. Counterexamples to a proof, animals that do not fit local taxonomies, or deviants who violate current norms all generate the same range of reactions.

Lakatos examines proofs and refutations around Euler's theorem (put forth in 1758) that for polyhedra the number of faces (F), edges (E), and vertices (V) are related by $V - E + F = 2$. The theorem is readily verified for cubes, prisms, and pyramids. The question is: does it work for all polyhedra? Lakatos says, in Bloor's words, that "the stability and scope of every theorem is precarious: critical argument and adjustment is in principle endless; there is no final truth to reveal, only a ramified and interlocking network of claims and counterclaims to be balanced and stabilized".[15] But Bloor goes further. "If", he writes, "the stream of potential counterexamples is endless then the processes whereby we accord or fail to accord recognition of them must also be endlessly at work". Lakatos has revealed mathematics as something that has to be *negotiated*. But Lakatos doesn't tell us what is at stake in these negotiations. For example: "What investments do people have in changing or in maintaining the boundaries of mathematical concepts"?[16]

Now when we find mathematicians engaged in monster-barring, Bloor notes, we recall the abominations of Leviticus (Lakatos has one of the characters in his dialogue say, "It is a *monster*, a pathological case, not a counterexample"; the small stellated dodecahedron, says Schafli, is not "a genuine polyhedron"). Mary Douglas reminds us that the abominations were a list of anomalies to the animal classification system in the Pentateuch. So monster-barring can be correlated with pollution activities; monster-adjustment and exception-barring can be correlated with ritual. It then becomes evident that if styles of ritual and pollution are correlated with types of social organization, we should be on the lookout for comparable types of social organization where the styles of monster-adjustment and monster-barring appear. In order to pursue this idea, Bloor suggests we turn to Mary Douglas' grid-group theory. Douglas and Wildavsky describe this theory as follows:[17]

> This is a way of checking characteristics of social organization with features of the beliefs and values of the people who are keeping the form of organization alive. *Group* means

the outside boundary that people have erected between themselves and the outside world. *Grid* means all the other social distinctions and delegations of authority that they use to limit how people behave to one another.

Earlier (Chapter 7), I discussed grid and group in terms of classificatory systems. Classificatory decisions are made with reference to social, natural, and symbolic worlds. The correspondence between a classificatory system and a given environment refers to *any viable relationship that can be established between them.*[18] The stability of a knowledge system or network as a whole and in its parts varies; the whole network and network parts are subject to *varying types and levels* of negotiation. This is something Bloor tends to overlook – he suggests that every part of the network is subject to the same type and level of negotiation. He also ignores power struggles and more militant forms of conflict as means of reaching classificatory decisions.

As I suggested in Chapter 7, the synthesis and application of interest, network, classification, and constructivist theories is the most promising theoretical task facing contemporary sociologists of knowledge. Bloor has taken us a long way on the path to completing this task. One of Bloor's colleagues, Donald MacKenzie, has also made an outstanding contribution to what I have referred to here as "Spenglerian" or "materialist" sociology of mathematics.

If there is any life left in the "Weissstorch theory of knowledge", MacKenzie's study of *Statistics in Britain 1865–1930* will help to extinguish it.[19] The stork didn't bring us Student's t-test, it was a product of Gossett's studies of beer behavior in a Guinness brewery. But statistics is not merely a social *product*; it is, MacKenzie argues, like all knowledge *constitutively social*. It is perhaps too extreme to conclude on this view that the tetrachoric correlation coefficient, for example, is racist and elitist because it served eugenic interests. But Pearson's correlation coefficients reflected his view of correlation as a measure of "the strength of heredity" and a tool for evolutionary and eugenic prediction. Earlier, Galton had developed regression and correlation in connection with *his* eugenic program. (Galton referred to eugenics as a "virile creed"; the intimate relationship between statistics and eugenics documented by MacKenzie suggests the label "virile science" for statistics. It is not incidental that the *Eugenics Review* regularly endorsed conventional notions of masculinity and femininity).

The error theorists who preceded Galton were not, MacKenzie persuasively contends, "on the verge" of a simultaneous discovery. The major reason is that the error theorists and Galton had different goals and different views of variability. The Belgian pioneer in statistical analysis, Quetelet, does not figure prominently in this account of statistics in Britain. It is interesting to note, however, that Quetelet (whose work Galton was familiar with) consistently stressed that statistical regularities – in crime rates, for example – did not reveal the forces of heredity but rather showed that society "contains the germs of all the crimes that will be committed, as well as the conditions under which they can develop". In his *Sur l'homme et le développement de ses facultés, essai d'une physique sociale* (1835), he wrote that his observations showed the possibility of treating social ills by modifying institutions, habits, and education. MacKenzie's view is that statistics are tools; and while a tool's construction will reflect the tasks it was designed for, the tool is not always limited to those tasks. Thus "our statistics" and "Pearson's statistics" must be evaluated in terms of their own contexts of construction, development, and use. More generally, MacKenzie views science as an activity of invention, oriented to the general goal of improving the human capacity for prediction and control. The pursuit of particular goals is sustained by social interests located either within science or the wider society. Societal interests can reflect the *overall* (e.g., class) or *fine* (e.g., social institutions) structure of society, or the relative influences of overall and fine structures.

MacKenzie places the lives and works of Galton, Pearson, and R. A. Fisher in their social contexts, and analyzes them as key points in the development of statistics as a scientific specialty. This development conforms to the Griffith–Mullins network model, a model MacKenzie suggests was anticipated by V. I. Lenin in his theory of the "party" and revolutionary change. Fisher's work, according to MacKenzie, extended the contributions of Galton and Pearson, his bitter disputes with Pearson notwithstanding. Fisher's biological research was clearly tied to his eugenic concerns. But the close relationship between statistics and eugenics evident in the works of Galton and Pearson is not found in Fisher's case. His "metastatistics" focused on refining the practices and theories of the earlier generation of statisticians. MacKenzie does not make the organizational basis of Fisher's metastatistics explicit enough. "Metastatistics" was just what we would expect from someone working in a field that had become a viable specialty with problems bequeathed directly from one professional generation to the next. The strengthening of that organizational imperative underlies Neyman's argument that "confidence limits" were more rigorous than Fisher's "fiducialism", a controversy that takes us beyond the 1930 boundary of MacKenzie's story. Fisher's work, of course, was not entirely divorced from extra-statistical concerns. MacKenzie notes that it was closely tied to agricultural research; it also reflected a new role for statisticians in scientific research (especially in the areas of agricultural and industrial production) that had begun to emerge in the 1920s.

According to MacKenzie, studies of science and ideology abated in reaction to the Nazis' "Aryan physics" and the Stalinists' "proletarian biology". He is intent on resurrecting such inquiries, but not in order to argue that modern science is "bourgeois", or that statistical theory is "fascist". Instead, he concentrates on how social interests influenced the form, content, and development of statistical theory. His general thesis is that advocacy of and (less frequently) opposition to eugenics motivated statistical work and affected the content of statistical knowledge during the period in which statistics developed out of Galton's work and Pearson's institution-building into a scientific specialty. Eugenics was one manifestation of the social interests of a rising professional middle class in late nineteenth-century British capitalist society. Eugenicists argued that their policies (which stopped short of the "lethal chamber" but included "segregation") were based on scientific naturalism, and more specifically on Darwinian evolutionary theory. MacKenzie contends that in fact the eugenicists "read" class structure onto nature. (It is ironic that Darwin, who is called on to do battle in "monkey trials" and to vanquish Creationists, was linked to the surrogate religion of eugenics (see below) by blood and argument. He gave inheritance the edge over environment in individual development; Galton was his cousin; and his son, Major Leonard Darwin, was President of the Eugenic Education Society from 1911 to 1929).

Much of post-Mertonian sociology of science has a neo-functionalist slant that ignores or obscures conflict, elitism, and issues of power and social control. MacKenzie, however, places conflict and power at the center of his story: biometricians led by Pearson versus Mendelians led by Bateson; Pearson versus Yule; Pearson versus Fisher; ideas used as weapons within science and in the relationships between science and society; scientists seeking dominance over other scientists; and scientists pitted against other professionals in an attempt to establish science as "the sole arbiter of rational belief". On the latter issue, I wonder that MacKenzie didn't pay more attention to the role of the British Association for the Advancement of Science, which supported some eugenics research and stressed "scientific method" in its public relations efforts on behalf of science and scientists. This would have, for one thing, rooted Pearson's emphasis on method and the internal struggles among statisticians over methods in an important organizational context.

The institution Galton and Pearson constructed became inevitably dominated by what MacKenzie, following Becker, refers to as "side bets", that is, other interests besides eugenics. MacKenzie's story has its own side bets: (a) reading his account of the eugenics movement in England prompts one to wonder about Nazism as an extreme version of a civilizational racism instead of a Teutonic aberration; (b) professionalism was so closely tied to eugenics that it seems reasonable to conjecture that it was an alternative eugenic strategy, a social genetics for the controlled selection, socialization, reproduction, and nourishment of a "perfected" elite. A third side bet, the relationship between statistics and religion, is especially intriguing. MacKenzie shows that the micropolitics of the biometrician-Mendelian controversy paralleled the decline of religion. Increasingly, intellectuals defended their social and political views on the authority of Nature, not of God. The opposing factions led by Pearson and Bateson constructed different biologies, and used them to defend different social arrangements. I view this as more a matter of a translation in religious space than one of the decline of religion. MacKenzie himself suggests a translation by noting that eugenics was a surrogate religion for Galton and Pearson. Nature, as I suggested earlier, replaced God. As early as 1738, DeMoivre attempted to free probability from its association with gambling, and to establish its theological relevance in a doctrine of a "divine order" exhibited in "the regularity of statistical ratios". MacKenzie in fact discusses "social Newtonianism" and the framework it provided for much of the development of probability theory in eighteenth-century England; Bayesianism, for example, has its origins in the quest by the Reverend Thomas Bayes "to confirm the argument taken from final causes for the existence of the Deity". (Pearson exhibited an awareness, if not necessarily a self-awareness, of the theological connection when he observed that statistics were a religion for Florence Nightingale; "the Passionate Statistician", as she was sometimes referred to, had been J. J. Sylvester's most distinguished mathematics student in the 1850s).

MacKenzie's book is a significant contribution to the constructivist perspective in social studies of science. MacKenzie clarifies, if he does not entirely resolve or dispose of, the problem of understanding knowledge as "constitutively social". Implicitly, he also provides resources for a psychological theory that treats knowledge as "constitutively affective". He uses sociological theory to penetrate about as close to the social heart of mathematical knowledge as anyone has managed to get (albeit in a case that is not so recalcitrant when compared to Greek geometry or Cantor's invention of transfinite numbers).

MacKenzie vacillates between determinism and voluntarism (while generally striving, rhetorically, to give voluntarism the edge), and (less emphatically) between description as pure and as inevitably theoretical. He defends bold conjectures, but he throws up so many qualifications that he transcends due caution and distracts himself and his reader from precisely the sort of daredevil theorizing we need if we are going to develop a critical instead of a worshipful understanding of science. And is there, after all, a role for truth and causality in this inquiry (and its constructivist and relativist relations) hidden by a soft determinism and a surrogate causal rhetoric? Finally, it is important that those of us who affirm MacKenzie's perspective and conclusions, as well as those who oppose them, consider whether and how MacKenzie's interpretation of the social construction of statistics has been influenced by his evaluation of statistical methods. This is a study that eschews the individualist, empiricist, statistical approach and adopts a sociology of knowledge perspective influenced by Lukács and Goldman. Following MacKenzie's lead, we can expect to solve some of these puzzles by examining his use of rhetorical resources to persuade readers (especially those readers who are not experts in statistics and who make up an important part of the intended audience for his book) about his viewpoints. And we can trace the interests that lie at the root of why he finds it "useful" to consider certain conjectures "reasonable".

CONCLUSION

The limitations of the Marxist and evolutionary perspectives on mathematics are unnecessary concessions to Mannheim's ghost. Spengler's approach is generally more tenable sociologically. For some scholars, Spengler's world view approach may raise the twin spectre of solipsism and relativism. But it is impossible to separate even "obvious" statements such as "$2 \times 2 = 4$" from the process of social construction and assert that they represent a universal, eternal, ultimate reality, or that they represent *the* one-and-only "true", "objective" reality as opposed, for example, to a Doestoevskian reality in which "$2 \times 2 = 5$".[20] Certainly, there is a "reality" that exists independently of the lives and deaths of individual human beings and the rise and fall of civilizations; but it is a reality human beings are part of, a reality that can only be known and sustained through the constructive activities of human beings, and a reality that is more or less common to us all to the extent that we participate in the same or overlapping communities of consensus.[21] But the novelty that arises at the intersection of unique individuals and cultures, and the universals of human experience, make it inevitable that such similarities or overlaps must eventually give ground to incommensurable differences.

APPENDIX D

THE SOCIAL ROOTS OF NON-EUCLIDEAN GEOMETRY

Non-Euclidean geometries (NEGs) have their origin in problems with Euclid's parallels postulate already noted by his earliest commentators. Later, the names of Saccheri (1667–1733), Lambert (1728–1777), Klügel (1739–1812), and Legendre (1752–1833) are among many that mark a continuous interest over the centuries in the parallels postulate. The first important social fact, then, is that there is a historical context for NEGs.

The best known creators of NEGs – Lobachevsky, J. Bolyai, and Riemann – were by no means provincial isolates. In fact, all were directly or indirectly linked to Gauss (1777–1855) at Göttingen. As early as the 1820s, Gauss had concluded privately that NEGs were possible. Gauss and Wolfgang Bolyai (1775–1856), Johann's father, were fellow students and friends at Göttingen. Abraham Kästner (1719–1800) was a professor at Göttingen during Gauss' and W. Bolyai's student days. Kästner had studied the many attempts made from the time of Euclid's commentators on to prove the parallels postulate, and he discussed this work in his lectures. There are letters on NEGs written by Gauss to W. Bolyai (December 17, 1799), to Taurinus (November 8, 1824), and to Bessel (January 27, 1829). Gauss' other remarks on NEGs are in the form of unpublished notes (1831) and two short reviews in the *Göttingische gelehrte Anzeigen* of 1816 and 1822.

The problem of parallels was certainly, as Struik has observed, "in the air" at Göttingen; dissertations were written on the problem. There were other students of the parallels postulate at Marbourg, such as F. K. Schweikart (1780–1859) and A. L. Gerling (1788–1864). Gerling had studied with Gauss. Schweikart's nephew Taurinus (1794–1874) had studied at Göttingen (though after Kästner's death).

Wolfgang Bolyai became a mathematics teacher in a provincial Hungarian town; but he devoted a great deal of energy to the parallels problem, and expressed dismay when he found that his son Johann (1802–1860) had become intrigued by this intractable problem. Johann's 'The Science of Absolute Space' appeared in a book written by his father in 1832–1833, but he had developed his ideas on NEGs as early as 1823.

Lobachevsky (1793–1856), like the Bolyais, seems to qualify as an "isolate" at first glance. However, the "provincial" university he studied at,

Kazan, was staffed by distinguished German professors, including his teacher J. M. Bartels (1769–1836). Bartels had studied with Kästner at Göttingen, and was one of Gauss' teachers. Lobachevsky was professor and rector at Kazan from 1827–1846. He presented his ideas on the foundations of geometry in numerous papers and two books from 1825 on.

Riemann (1826–1866), in spite of modest origins, studied under Gauss at Göttingen (after some work at Berlin), and became *Privadozent* there in 1854. His *Habilitationschrift*, written under Gauss' tutelage, was titled "Über die Hypothesen welche der Geometrie zu Grunde liegen".

Clearly, Lobachevsky, Bolyai, and Riemann were members of a network of *some* sort. And while this leaves many questions about mutual influences open, it certainly makes some accounts of the development of NEGs look curious. Even while documenting the historical and social contexts of NEGs, Boyer and Struik still express "wonder":[22]

> ... we find a startling case of simultaneity of discovery, for similar notions occurred, during the first third of the nineteenth century, to three men, one German, one Hungarian, and one Russian (Boyer).
>
> It is remarkable how the new ideas sprang up independently in Göttingen, Budapest, and Kazan, and in the same period after an incubation period of two thousand years. It is also remarkable how they matured partly outside the geographical periphery of the world of mathematical research (Struik).

Now there should be no mystery if people *working* more or less independently bring together similar cultural resources and universal cognitive propensities and abilities conditioned by common educational experiences, and arrive at similar conclusions. More importantly, it is not generally recognized that NEGs were created to *solve problems* that had arisen in mathematics and in studies of physical phenomenon. Lobachevsky referred to his geometry initially as "imaginary" not because it was a pure mental or logical construct based on arbitrary premises but because he had not yet formulated a clear physical explanation for it. Geometrical ideas, he wrote in 1835, "can only be verified like other physical laws by experiments, such as astronomical observations". He considered the properties of space to be dependent on matter, and asserted that the laws of gravity might change in a transition to a new physical domain.[23]

One of the consequences of the emergence of new geometric ideas in the works of Lobachevsky and others was the creation of the geometry of many-dimensional spaces. This work was motivated in part by the geometrical approaches to solving problems in algebra and analysis. If there is a need to mystify mathematics, then it is easy enough to make the idea of a many-dimensional space mysterious and mystical. But an n-dimensional space is simply a context for n points representing n data on some figure, system state, etc. By using this representation (as in the case of ordinary graphs) it becomes possible to utilize well-known geometrical methods and analogies to solve problems. Aleksandrov writes:[24]

> ... there are such things as a sphere or a mixture of three gases that are characterized by several data so that the collection of all these things is many-dimensional. The number of variables in a given case is just the number of these data: just as a point moving in space changes its three coordinates, so a sphere moving, expanding, and contracting changes its four "coordinates", i.e., the four quantities that determine it.

Mathematicians use "space" in a general sense to refer to "an arbitrary collection of homogeneous objects (events, states, functions, figures, values of variables, etc.) between which there are relationships similar to the usual spatial relations (continuity, distance, etc.)". Objects become "points" of such a space; figures become sets of the space's points.

Riemann treated geometry in a very general way in his 1854 lecture. There were no calculations and no proofs. However, the "formal apparatus of his theory" appeared in another paper, applied to the problem of heat conduction. Riemannian geometry provides an analytical expression for the difference between a general Riemannian space and a Euclidean one. The often mystified idea of "space curvature" is simply a measure of the "non-Euclideaness", in a sense, of a Riemannian space:[25]

> ... it is defined intrinsically as a measure of the deviation of its metric from the metric of Euclidean space. It determines, for example, the difference of the sum of the angles of a triangle from π or the difference of the length of a circle from $2\pi r$.

Thus, Riemannian geometry reflects the reality that the real properties of space may differ more or less from what Euclidean geometry states.

Now all of what I have rehearsed here may seem trivial to historians – once they have uncovered the connections and more importantly recognized their significance, perhaps, for a sociology of mathematics. My point is not that everything about this case is crystal clear but that information gaps and interpretative lacunae notwithstanding we find a situation that is consistent with the sociologist's expectations. The *social* history of NEGs, unencumbered by elaborate sociological theories, is enough to undermine tales of abstract mathematics as an enterprise for heroes and magicians. This is only a small step for the sociologist who wants to develop a systematic explanatory account.

In the next two chapters, I sketch a materialist sociological account of mathematics that affirms the legacy of Marx reinforced by Spengler's promptings toward a consistent materialism.

SECTION VIII

SOCIOLOGICAL MATERIALISM AND HISTORY OF MATHEMATICS: AN EXPLORATORY CASE STUDY

CHAPTER 14

MATHEMATICS IN ANCIENT GREECE

INTRODUCTION

In this and the following chapter, I outline a materialist sociological account of mathematics in ancient Greece from Thales to Archimedes, and in Europe from Stevin and Valerio to Newton and Leibniz. There are two reasons for examining the Greek and European cases together. One is that Alexandrian mathematics reached the threshold of pre-modern European mathematics. Stevin, Valerio, Guldin, and Cavalieri begin their mathematical labors in the sixteenth and seventeenth centuries essentially where Archimedes' work ends. The major thematic link between the two periods is the calculus.[1] The second reason is that there are significant similarities between the social changes associated with the transition from the Middle Ages to the Renaissance in Europe and the transition from the "Greek Dark Ages" to the Ionian commercial and cultural awakening in the seventh century BC.[2] In a sense, the Europeans "recapitulated" significant aspects of Greek mathematical, social, economic, and political history in a way that allowed them to take advantage of and build on the Alexandrian legacy. I extend earlier materialistic studies of mathematics in these two periods by (a) focusing on the continuities between ancient Greek and pre-modern European mathematics, (b) linking "social development, industry and commercial intercourse"[3] to the development of mathematics and mathematical communities, (c) identifying the activities and products of communities of mathematicians as material bases for mathematical development, (d) eliminating idealist and transcendentalist assumptions in accounting for "abstract" or "pure" mathematics.

ASSUMPTIONS AND CONJECTURES: I – MATHEMATICS AND SOCIETY

Marx and Engels argue that even the simplest sensuous certainties are products of "social development, industry and commercial intercourse". I interpret this in a strong sense; there is no "external reality" *an und für sich*. All knowledge is constituted of mental-and-physical activities, culture, and history. There is no a priori reason to exempt mathematics from this premise.[4]

The products of mathematical work range from mathematical ideas and theories to mathematics treatises and textbooks. Materialist sociology of mathematics focuses on the relationship between the mode(s) of production (first order relationships) and the social organization of mathematics (second order relationships), and the nature and products of mathematical work. Mathematics is neither transcendental (something "out there" in an eternal, universal realm of reality that we can reveal in one way or another), nor idealist (a product of "pure" mental activity), nor an heroic enterprise (the work of individuals, sometimes labeled "geniuses", who create it virtually "out of nothing").

Mathematics in any given social formation grows out of practical activity and corresponds to the prevailing mode(s) of production, distribution, and consumption. But mathematics is not a simple "reflex" of a narrowly conceived "economic base". It is a product of interactions among various activities and products within a social formation. The more complex the social formation is, the more complex the process of constructing knowledge is. The study of the complex social construction of mathematics, and the development of a coherent theoretical account of mathematics and society fall outside the aims of this analysis. Indeed, the groundwork for such a project has not yet been completed. I limit myself, therefore, to an exploration of mathematics as a product of social development, industry and commercial intercourse in the most general terms.

ASSUMPTIONS AND CONJECTURES: II – SOCIAL, ECONOMIC, AND POLITICAL FACTORS

Karp and I have hypothesized elsewhere that economic decentralization and exchange, and political decentralization are interrelated and mutually facilitative. Together, they constitute necessary conditions for the functional differentiation of more or less institutionally autonomous activities. The details of, rationale for, and substantive illustrations of this hypothesis are given in our study of ecological factors in the emergence of modern science.[5] On the basis of this earlier work, I view the emergence of commercial activity (including commodity exchange in seventh century BC Ionia, the development of decentralized cities around "mother-cities" such as Miletus, and the development of the Greek city-states) as factors in the functional differentiation of more or less autonomous social activities in ancient Greece. These conditions, for example, are associated with the emergence of a merchant class, and schools of philosophy.

The new social, economic, and political conditions in seventh century BC Ionia supported the development of a network of philosophers (i.e., the School of Miletus, and more broadly the Ionic School). During the Ionic period, theory and basic sustenance labor were not separated. Reason was applied to problems of trade and commerce, and work was not despised by the emerging merchant class. However, a slave trade emerged on the coastal island of Chios, located about one hundred miles northwest of Miletus. From 600 to 400 BC, a steadily increasing part of the production process and the administration of labor was transferred to slaves. An elite philosopher class, growing out of the Ionian merchant class, became increasingly divorced from the everyday world of work. The divorce of philosophical from basic sustenance activities is often referred to as "the separation of hand and brain". This sort of separation is found in other societies, but under different social, economic, and political conditions. The extreme separation of hand and brain under Platonist oligarchies in classical Greece exhibits some similarities to the separation under Oriental Despotism. However, the Greek separation occurred against the background and continuing influence of the Ionian commercial revolution. The level of mercantile activity, and the degree of political and economic decentralization in ancient Greece, although variable, distinguishes Greek civilization from, for example, Chinese civilization. This distinction helps account for the difference between Greek and other (e.g., Chinese) elites.

A specialist class of philosophers emerged alongside and close to the oligarchic political and economic leaders who came to power in the classical period. Their teacher-student networks extended over numerous generations. With each generation, their symbolic systems became more general, more abstract. New generations of philosophers built their ideas on the foundations of earlier symbolic systems. These symbolic systems served as the material bases for the productive labor of philosophers linked in teacher-student networks and working in itinerant or formal schools.

Symbols are resources which human beings react to, create, and manipulate in their struggles for subsistence and for social, economic, and political development. They are a different sort of material reality than *natural* and *human-made technological* resources. Societies and cultures are crucibles for the creation of new levels of material reality, including symbolic systems, and therefore of new foundations of human awareness and activity. Furthermore, symbolic systems are created by people working in certain types of social groups. Therefore, symbols and the rules for manipulating them can be expected to reflect the organization and activities of the groups that create them; they *embody* social forms, activities, and processes. A detailed discussion of the embodiment hypothesis cannot be presented here. However, clarifying the relationship between symbols and social organization in terms of the embodiment hypothesis may be the key to fully eliminating idealist and transcendentalist assumptions from materialist studies of abstract mathematics.

As hand and brain were separated in ancient Greece between 600 and 400 BC, the generalizations of the philosophers were increasingly "purified" of basic sustenance labor and practical demonstration. The degree of "purification" or abstraction was more limited than generally recognized, and exaggerated for ideological reasons by the Greeks themselves as well as later admirers of Greek mathematics.

Abstraction depends on realizing opportunities for producing, publishing, and disseminating ideas in a specialized community of teachers and students that extends over a number of generations. If the abstracting process is carried out under conditions of social fragmentation (e.g., class divisions and class struggles) and the fragmentation of knowledge (e.g., separating and opposing natural knowledge and political-ethical knowledge), it will tend to generate ideologies of purity. This is not simply a consequence of "professional autonomy" and specialization; it is an imperative of class struggle. The "thinking class" in classical Greece had the time but also the charge from their oligarchic leaders (however much this may have been a latent product of socialization) to reflect on and elaborate numbers. The philosopher-mathematicians from Thales to Archimedes were worldly people, aware of and active in political, religious, military, and economic affairs. In great part, their work consisted of rationalizing thought and developing algorithms for the ready application of ideas to political, military, and economic ends. But the more insulated their work became from "external" interests, the more the algorithmic imperative was subordinated to the imperatives of teacher-student relationships and traditions, and to the direct or indirect support of ruling class ideas and activities. The self-consciousness about and the defense of professional insularity led to the development of ideologies that justified and glorified the separation of hand and brain; this also reflected the role of the thinking class as defenders of a slave-based society. Ideologies of purity are thus associated with the extreme separation of hand and brain in classical Greece. They did not develop among the Ionians. The classical ideology retained some of its strength even as hand and brain were reunited in Alexandrian Greece.

Theory is related to the speculative thought that accompanies the division of society into classes under conditions such as those which developed in ancient Greece. It is, in general, the special possession of the ruling classes. As a ruling class becomes divorced from basic sustenance labor, its thought patterns become increasingly "speculative". These thought patterns correspond to the relatively isolated and insular material conditions of the ruling class style of life. A scholarly subclass within or supported by the ruling class transforms speculation into theory.[6] This sub-class is directly or indirectly dependent on the patronage of the ruling class. The distinction between speculation and theory depends in part on the degree to which the theory specialists can and do maintain some ties to productive labor. To the extent, however, that this process is based on and reinforces the separation of hand and brain, it will undermine itself. Already with Aristotle, there was some difficulty in understanding earlier authors because the bases of their work were too far removed from common experience, discarded on account of a distaste for demonstration, or overspecialized. Post-Euclidean scholars had comparable problems in interpreting earlier mathematical works.

The development of a more or less autonomous institutional sphere creates a new context for productive activity. The same sort of analysis that at a lower level begins with activities aimed at satisfying basic human needs must now be carried out simultaneously on the new level where one aim is to satisfy organizational and institutional imperatives. The institutionalization of mathematics creates the conditions for generating mathematics out of mathematics. This does not mean that the material foundations of mathematics are transcended.[7] Rather, they become more complicated; new organizational levels are built upon lower level sustenance organizations. Occupations, professions, and institutions are higher order material foundations for human productive activity.

What was missing from the non-Greek world was a certain degree of functional differentiation of interrelated and interdependent, institutionally autonomous and specialized activities. The impetus for the development of Greek mathematics (as well as of science, and more generally "culture") came, especially in the Ionian period, from communication, competition, and conflict among politically and economically diversified communities. The other factor important in the development of Greek mathematics is generational continuity. In particular, it was the *scale* and the *continuity* of the features I have identified as crucial in Greek social organization that distinguished Greece from other early civilizations. It is important to recognize that these features of the Greek social landscape were limited, and changed during the period between 600 and 300 BC. Diversification and exchange, for example, are already in decline in the classical period.

A MATERIALIST ACCOUNT OF MATHEMATICS IN ANCIENT GREECE[8]

Farrington's study of Greek science is designed to show that science originates in the various activities people engage in to sustain themselves physically and spiritually. He argues that science arises out of human experiences with things, is dependent on sensory evidence, and develops in step with the stages of social progress. The history of Greek mathematics affirms Farrington's historical materialist argument.[9]

"Greek mathematics" actually refers to Greek geometry. The basic features of Greek geometry are "pure reason", "deduction", and the manipulation of "abstract universals". These features are to some extent the creations of students of Greek mathematics and culture. Out of the context of Greek culture, and unqualified, the adjectives "pure", "deductive", and "abstract" often obscure more than they reveal about Greek mathematics. Nonetheless, they serve to identify the mathematics created by the Greek philosopher-mathematicians from Thales to Archimedes. To be more accurate, however, Greek mathematics should be identified with the philosopher-mathematicians of classical (post-Ionian) Greece. Ionian mathematics was not as far removed from the arithmetic widely used in household and commercial transactions as was classical mathematics. Still, the development of Greek mathematics has its roots in the Ionian and Pythagorean schools. In many respects, classical Greek mathematics is a systematization and rationalization of Ionian and Pythagorean achievements. The Pythagoreans, for example, had achieved high levels of abstraction.

The Ionians. The social, economic, and political conditions necessary for the functional differentiation of diversified, autonomous, and interdependent activities such as philosophy, astronomy, and mathematics emerged in Ionia beginning in the eighth century BC. There was a renewal of contact with the Near East, the establishment of politically independent Greek colonies, and the gradual development of a limited exchange economy between and among the parent state and the colonies. The limited accumulation of capital provided resources for the patronage which was an important factor in the development and continuity of the philosopher class. Between 750 and 500 BC, then, new cultural contacts, political and economic decentralization, and the development of limited commodity exchange established the context for Ionian and Pythagorean achievements in mathematics.[10]

In the eighth century BC, contact between the Ionian cities and the kingdoms of Phrygia and Lydia was established. Lydia was an active commercial center on the overland trade route between Babylonia-Assyria and its neighbors. There was a road connecting Lardes, the capital of Lydia, and the Assyrian city of Nineveh. The confluence of trade and commerce in Lydia must have facilitated the development of a medium of exchange. Confluence alone, however, cannot account for the apparent invention of coinage in Lydia. Ederer suggests that the selection of the abundant metal electrum as a monetary medium led to difficulties.[11] Variable in content and color, electrum was not a good choice for monetary purposes. In any case, it appears that apart from Greece and a few isolated cases only India and China among the ancient civilizations developed a native coinage. It is likely, however, that coinage diffused to China and India from the eastern Mediterranean. Coinage was a sporadic phenomenon outside of Greece; and only in Greece were coinage and exchange associated on a significant scale and over an appreciable span of time.

The commercial revolution in Ionia brought great merchant families to power. Miletus, the coastal mother-city, was the leading political and cultural center in Ionia, and the center of the Oriental trade network. It was the birthplace of Thales (ca. 640–546 BC) and his students Anaximander (ca. 610–547 BC) and Anaximenes (fl. 545 BC). Reconstructions of the lives of these Milesians are based on fragmentary evidence. Nonetheless, what has come down to us about the Milesian philosopher-merchants at least symbolizes the Ionian achievements in commerce, science, and mathematics.

Thales was probably a member of the new merchant class. He and his followers were influenced by and helped to fashion an increasingly well-organized economy. Their commercial interests led them to seek an improved understanding of the natural world and of computational methods. Their methods of inquiry were probably closer to those of the earlier poets than to those of the later classical philosophers. It is likely that reasoned argument was still subordinated to simple assertions in seventh century BC Ionia.[12] But commercial imperatives stimulated rational calculation. Thales, in fact, is sometimes credited with introducing the mathematical proof. He is supposed to have proved the proposition, known from earlier times, that a circle is bisected by its diameters. His method of proof must have been an extension (generalization) and articulation of earlier methods. It was probably more advanced that the Babylonian notion of proof which involved little more than "checking" results, and the knowledge that certain procedures would always give certain results[13]. The Babylonians did not articulate their proofs in theorems. Thales may have made some progress in this direction.[14] The rationalization of such efforts by later philosopher-mathematicians eventually culminated in Euclidean-type proofs.

Milesian "philosophy" was more a matter of rational calculation about the natural world in the interest of commercial exploitation than a search for "wisdom". The Greek word *Sophia*, for example, often taken to mean wisdom and abstract thought, meant "technical skill" in the pre-Socratic period, and may have retained this meaning to an even later date.[15] In classical philosophy, it should be noted, the search for "wisdom" was partly an ideological cloak for the intellectual activities of ruling class apologists.

The Ionian achievements in mathematics are not miraculous. There are clear continuities between lonian mathematics and the mathematics of Babylonia and Egypt. The Greeks themselves report that Thales traveled in Babylonia and Egypt. If Thales did in fact predict the eclipse of 585 BC he did so with a knowledge of Babylonian astronomy. His travels and the results thereof illustrate the connection between mathematical and mercantile activities. In the process of pursuing economic interests and simultaneouly searching for improved methods of rational calculation, Thales and later Greeks encountered the mathematics of Babylonia and Egypt. That mathematics, like all pre-modern "autochthonous" mathematics (excepting classical Greek mathematics) was primarily inductive; it was a "mathematics of survival". Some Babylonians and Egyptians were able to systematize and generalize the common experiences of earlier generations with weights and measures. There was ample opportunity for the Ionian and later Greeks to learn about these accomplishments. There is documentary evidence, for example, indicating that the Babylonian scientific tradition was carried on without serious interruption well into the Hellenistic era.[16]

The theoretical achievements of the Babylonians and Egyptians in mathematics are often underestimated. It is certianly true that theory was not well-developed in these civilizations. But Egyptian documents, for example, contain systematic methods for treating relations between fractions, the method of false assumption (London leather-scroll), and (in the Moscow papyrus) the correct calculation of the volume of the frustrum of a pyramid with square base using the formula $v = (a^2 - ab + b^2)(h/3)$. The systematic approaches to problem-solving in the Rhind papyrus suggest that they were used in schools for scribes. These examples indicate that the Egyptians engaged in the types of systematizing and generalizing activities that are preconditions for the development of theories. The Babylonians were even more advanced in this respect.

Specialization among and continuity across the generations of philosopher-mathematicians who followed Thales made the development of increasingly general and abstract formulations of mathematical ideas possible. Proclus' (fifth century AD) history of geometry from Thales to Euclid, written as a commentary on the first book of Euclid's *Elements*, sketches a network of teachers and students stretching over a 300 year period (600–300 BC). In the second century BC, Hipparchus expressed an awareness of the importance of generational continuity in making possible the refined conclusion about the precession of the equinoxes. He purposefully addressed himself to recording data on the fixed stars which future astronomers could check in order to discover changes.

Social opportunities for philosophical and mathematical pursuits, and generational continuity, made it possible for the Greeks to carry the systematization and rationalization of centuries of human experience started in earlier times to new levels. At least initially, the separation of hand and brain that divided slaves, citizens, and political, economic, and intellectual elites facilitated the development of Greek science and mathematics. But the philosopher-mathematicians lost touch, to some extent, with basic sustenance labor, and this affected the nature of their generalizations and abstractions. At the same time, however, they developed enough political and economic power on their own and through patronage to enable their schools to last for generations. The separation of theory and basic sustenance labor reached its height during the time of Plato, but it was already in evidence among the Pythagoreans.

THE SEPARATION OF HAND AND BRAIN[17]

Pythagoras (fl. 532 BC) was born in Samos, a coastal island about thirty miles northwest of Miletus. He apparently associated (perhaps as a student) with Thales before migrating to Italy. Like Thales, he was probably a member of the new merchant class; indeed, he is credited with introducing coinage into Kroton.[18] The migration of Pythagoras to Italy is associated with the shifting of the center of Greek intellectual life from Ionia to the Greek colonies in Italy and Sicily.

The Pythagorean encounter with the irrational led to the realization that arithmetic could not be rationalized easily, if at all. Geometry, by contrast, lent itself to proofs and secrecy in ways that arithmetic did not. Arithmetic, as I noted earlier, was widely used in household and commercial activities. One indication of why the Greek philosopher-mathematicians preferred geometry over arithmetic is given by Florus in his comments at the dinner-table conversation reported by Plutarch:[19]

> For arithmetic, by its employment of number, distributes things equally, geometry, by the employment of proportion, distributes things according to merit. Geometry is therefore not a source of confusion in the State, but has in it a notable principle of distinction between good men and bad, who are awarded their portions not by weight or lot, but by the differences between vice and virtue. This, the geometrical, is the system of proportion which God applies to affairs.

The "incommensurables crisis" was resolved *geometrically* by Eudoxus (ca. 408–355 BC). His theory of proportion superseded Pythagorean arithmetical theory and made the notions of commensurable and incommensurable superfluous.[20]

The equation of Spartan oligarchy, geometry, and the law of God must have facilitated the development of the idea that the everyday world of the senses and the work of the "hand" were deceptive, and thereby stimulated the pursuit of an "axiomatic" geometry based on "reason".[21] This orientation to a highly rationalized mathematics could neither have arisen nor developed without specialization and generational continuity. But beyond this, the unique features of Greek mathematics reflected the separation of hand and brain.

The divorce of geometry from arithmetic (a contrast to the geometric algebra of the Babylonians) served at once to separate and to reflect the separation of social classes in Greece. On account of this development, the demand for a practical number system was never very great in Greece. Number symbolism not only failed to advance in Greece, it actually retrogressed relative to Babylonian achievements.

As the separation between hand and brain, between basic sustenance labor and elite philosophy, progressed from the Pythagoreans onward, the unique "purity" and "abstractness" of Greek mathematics progressed apace. But economic diversity and exchange never reached the levels they were to reach in Europe from the eighteenth century onward; so there were correlative limitations on specialization and professionalization. The levels of "purity" and "abstractness" achieved by the Greeks fell short of what the Europeans eventually achieved. Thus, the "purity" and "abstractness" of Greek mathematics does not really appear before Eudoxus (ca. 350 BC), a contemporary of Plato. The so-called "axioms" of Aristotle and Euclid were in fact "common notions" based on the common experiences of human beings over millennia.[22] Even as late as the fourth century BC, Aristotle was consciously systematizing human experiences. His "rules of logic" are, as he himself tells us, codifications of regularities he found in the ways earlier philosophers and mathematicians reasoned. Most of his examples are drawn from contemporary texts which focus on practical problems. Euclid (ca. 330–275 BC) may have had access to Aristotle's work; he was, in any case, well-acquainted with the Platonist geometry.[23] His "common notions" parallel Aristotle's codification. In general, the privileges of their social position made it possible for the Greek philosopher-mathematicians to carry out the codification of human experience; this was also an *imperative* of their social position, to the extent that their work served the interests of the state in the sense noted above by Florus.

The significance of the generational continuity of a specialized class as a precondition for the development of high levels of abstraction is indicated by the fact that prior to Plato, mathematical concepts are rarely if ever treated as "pure" abstractions. Aristotle himself notes that the early Pythagoreans, far from considering numbers as "transcendental", thought of them as the ultimate elements of material things. Their notion of number as the First Principle was tied to the political, economic, and religious interests of their aristocratic secret society. Their concept of number reflected their concept of and role in society. This is partly manifested in the stories about the discovery of incommensurable magnitudes, the attempts to keep the discovery a secret, and the disasters that befell those who sought to divulge the secret (e.g., the drowning of Hippasus of Metapontium). The problem of incommensurables, as I noted earlier, is related to the preference for geometry among the elite philosopher-mathematicians.

In the Pythagorean period, *Ionian* scientific and mathematical achievements began to wither, largely as a consequence of the divorce between hand and brain in a developing economy. Increasingly, the philosophers turned their attention away from the natural world, the focus of the Ionian philosopher-merchants, and toward the polis, religion, and the political and ethical justification of oligarchic power. The conflict between Ionian natural philosophy and the Platonist control of society through the "Noble Lie" is exemplified by the expulsion of Anaxagoras (ca. 500–428 BC) from Athens.[24]

Anaxagoras was born in Miletus, studied under Anaximenes, migrated to the port city of Elea in western Italy, and became the first philosopher to reside in Athens. Plato was born the year Anaxagoras died. Farrington writes that in the interval separating the active professional lives of the two philosophers,

. . . the attitude of Athens to Ionian science had become more clearly defined and the antagonism had deepened. It was not only that Socrates had begun his powerful movement of revolt against Ionian materialism; the technique of government through religion was also better understood as well as the threat to this technique inherent in the spread of Ionian nationalism.[25]

Anaxagoras still held the Ionian philosopher-merchant viewpoint that wisdom was achieved through "the possession of a capable pair of hands".[26] Bukharin refers to him as "the official state philosopher of Athens".[27] The expulsion of Anaxagoras conveniently marks a transition from Athenian commercial democracy (the Ionian legacy) to Platonist oligarchism.

In Plato, the separation of Thing and Idea reaches an extreme. This extreme separation corresponds to the extreme separation of hand and brain in Plato's Athens. By this time, much of the responsibility for and technology of basic sustenance production was in the hands of slaves. The elite philosopher class was virtually barred from activities in which nature was bent to human will under an economic imperative. Farrington writes:[28]

The [Ionian] conception of Nature as infinitely various and ingenious but inexorable in its laws is the conception of technicians who attempt to exercise over matter an operational control. The new [Platonist] conception of Nature, as a power with ends in view, which enforces its will on a subordinate but refractory matter, is the conception of a master who governs slaves.

Archimedes. Aristotle began to bridge the separation between theory and practice. By the time of the first Alexandrian school, hand and brain were being reunited. This was a tentative and fragile unity in comparison to the Ionian situation. It reflected political and economic perturbations in Mediterranean civilization that postponed but did not deter the withering of the Greek commercial impulse, and in particular of the remnants of the diversified exchange economy which had begun to develop in Ionia. Slavery was in decline, and experimentation (pursued with some sophistication among the Pythagoreans but employed sparingly thereafter) blossomed briefly. Given some appreciation of all this, it is somewhat easier to understand Archimedes' combined talents in engineering and "pure" mathematics, his apparent *distaste* for the "practical", his role as the last breath of Greek mathematics, and his death (at the hands of a Roman soldier) on the threshold of pre-modern European mathematics.

Archimedes' treatment of plane equilibrium serves as a microcosm of the generalizing aspect of Greek mathematical development. In Book I of his *Treatise on Plane Equilibrium*, Archimedes begins by establishing a set of postulates. These postulates are generalizations based on centuries of human experience with weights and balances: for example, equal weights at equal distances balance; if unequal weights operate at equal distances, the larger weighs down the smaller. Such postulates are the basis for arguments that lead to fundamental theorems. For example, based on the preceding postulates, Archimedes arrives at a theorem for *commensurable* magnitudes. Then, by *reductio ad absurdum* he *generalizes* this to include *incommensurable* magnitudes. The result is the following proposition: two magnitudes, *whether commensurable or incommensurable*, balance at distances reciprocally proportional to the magnitudes.

Archimedes was the son of an astronomer (Pheidias), and an intimate (perhaps even a relative) of King Hieron.[29] His *achieved* social position, at least, provided him with the opportunity to reflect on and elaborate common experiences, and to benefit from the generalizing activities of earlier generations of philosopher-mathematicians. As a consequence, he was able to arrive at a statement that appears to be "independent" of specific facts, paralleling the apparent independence of Archimedes' work from the vulgar interests of basic sustenance labor.

According to Plutarch, Archimedes "placed his whole ambition in those speculations in whose beauty and subtlety there is no admixture of the common needs of life". Plutarch, of course, was writing nearly three hundred years after Archimedes' death. In support of Plutarch's portrait, Heath points out that Archimedes wrote only *one* "mechanical" treatise, the work *On Sphere-Making*. But Heath here equates "mechanical" with "construction", for the lost treatise on sphere-making deals with the construction of a sphere representing motions of bodies in the heavenly system. Construction, however, is not the only sort of mechanical interest that can be opposed to "pure contemplation". In other works (some lost), Archimedes attends to common concerns such as the calendar, optics, centers of gravity, balances, and levers. The fact that he was an inventor (the greatest in antiquity according to some scholars), and that some of his inventions were designed for political and military purposes cannot (Plutarch, Heath, and Archimedes himself notwithstanding) be cast aside as "incidental", as Heath would have it. It is unrealistic to suppose that Archimedes could utterly detach his mechanical interests and talents from his interests and talents in so-called "pure mathematics". The co-existence of these interests and talents, amply documented, is sufficient grounds for arguing that Archimedes' mathematics was not a product of "pure contemplation". In his book on *Method*, discovered in 1906, he sets forth the mechanical bases of his formal ("pure") geometric proofs. His tendency to suppress the "vulgar" roots of the results he set out in a logical format for public consumption is reminiscent of some later mathematicians.[30] Archimedes' social position and the fact that he was able to generalize the generalizations of earlier generations and then to generalize his own could certainly have led him and admiring biographers (ulterior motives aside) to emphasize that his inquiries were not prompted by "vulgar" considerations.

Summary. "Greek mathematics" refers to the mathematics developed during the classical period of ancient Greek history, and especially to the systematization and rationalization of geometry. In contrast to the "Greek Miracle" idea, and Sohn-Rethel's exchange abstraction hypothesis, I have argued that (a) Greek mathematics was created on the foundations of earlier mathematical developments in Babylonia and Egypt, (b) this was possible because of favorable social, economic, and political conditions which influenced mathematical developments, and (c) among the most important factors which have to be taken into account in explaining the emergence and development of Greek mathematics are political and economic diversity and decentralization, specialization, and generational continuity. The Ionian and Pythagorean philosophers actually made the most important mathematical discoveries during a period of diversified and decentralized economic and political growth. The classical philosopher-mathematicians "purified" the earlier mathematical achievements. One result was the logical system of Euclid. The separation of hand and brain, however, reached an extreme in Plato's time. This, in conjunction with the rise of political oligarchies, resulted in a tendency toward an aristocratic treatment of mathematics. Unlike China, where the separation of hand and brain was continuously associated with Oriental Despotism over a long period of time, classical Greece built on the products of an earlier period of political and economic diversity and decentralization. The renewal of cross-societal social, economic, and political contacts in the Alexandrian period contributed to the unification of hand and brain to a certain extent, and stimulated a burst of creative mathematical activity. But the Alexandrians could not pass the threshold to pre-modern European mathematics.

CHAPTER 15

MATHEMATICS IN EUROPE, 1200–1700

The Legacy of Archimedes. The problem of passing from commensurable to incommensurable magnitudes led eventually to the development of the method of exhaustion. Here the Greeks took steps toward the creation of the calculus.[1] Already Antiphon in the fifth century BC had worked on "exhausting" the area between a circle and an inscribed regular polygon. This is a crude approach to integration. The fullest development of this method by the Greeks and the nearest approach to integration as we understand it is credited to Archimedes.

Infinitesimal magnitudes are encountered in the problem of motion, expressed in the famous paradoxes of Zeno of Elea. Problems of motion were crucial in post-Eleatic philosophy and are directly related to the Greek contributions to the calculus. These developments were linked to the military and commercial problems posed for the philosopher-mathematicians.

The comprehensive treatise on mechanics by Philo of Byzantium (Alexandrian school) illustrates what is meant by the statement that the Alexandrian Greeks reached the threshold of modern mathematics. The problems dealt with in this treatise are primarily military; industrial applications are notably absent. But this is probably because of the decline of commercial activity. In any case, the treatment of ballistics, and the clear display of experimental and empirical methods readily call to mind the problems and approaches associated with the development of the calculus in later European society.[2] In the European case, of course, the military problems were linked with economic ones in an increasingly active commercial environment.

Archimedes had invented weight-throwing engines, but the study of ballistics is notably absent from his work. Farrington attributes this neglect to Archimedes' "wholesome . . . awareness of the logical difficulties that beset the idea of motion".[3] This may be taken as yet another expression of the fact that the problem of motion had arisen for the Greeks *and* that it had outstripped the individual and cultural problem-solving resources of the time. The break in continuity created by the decline of Greek civilization meant that this barrier could not be broken through until and unless the continuity could be re-established.

The decline of Greece and the rise of Roman imperialism stifled mathematical work.[4] The next significant developments in the history of the calculus as a continuous enterprise are Stevin's *De Beghinselen der Weeghconst* of 1586, and Valerio's *De Quadratura Parabolae* of 1606. I do not intend my orientation here to be understood as a commitment to some sort of "continuity theory of history"; the situation is simply one of *cultural* development characterized by one or more "breaks". (It is important, of course, to recognize that cultural linkages across time and space allow items to pass from one *context* to another). Nor do I mean to minimize Arabic-Islamic influences and contributions. The continuity I am drawing attention to is expressed in the naming of Stevin as "the Archimedes of the Low Countries" and Valerio as "the new Archimedes". Cavalieri's Archimedean work on "indivisibles" was announced in 1629 and published in 1635. Stevin addressed his works to the practical needs and interests of working people, whereas Archimedes expressed contempt for "useful" science. Archimedes' expression of contempt, Farrington argues, signalled the "decadence" of Greek civilization.[5] By contrast, Stevin's perspective is, in a sense, "Ionian"; it is associated with and consciously tied to commercial expansion. In general, European mathematics in this period develops around the same general problem of motion that was at the center of post-Eleatic developments in Greek mathematics.

The development of the calculus from Stevin onwards is associated with the growth of exchange-value as an end-in-itself. Exchange-value did not fully mature in ancient Greece.[6] The fuller development of exchange-value in Europe was associated with a fuller development of diverse and specialized social activities. A mathematical community slowly coalesced, and, on the basis of ever-improving communication linkages forged in commercial transactions, experienced unprecedented growth. The professionalization of mathematical work was well under way by the nineteenth century. Under these conditions, mathematical abstraction was carried to levels only hinted at in the Greek achievements. But because abstract mathematics developed in conjunction with the development of exchange-value as an end-in-itself (the capitalist imperative), new levels of abstraction generally meant new levels of alienation within the mathematical community, in the relations between mathematicians and the natural world, and in the relations between mathematicians and other workers. Mathematicians and their clients and patrons transformed mathematical ideas into universal and eternal truths. An ideology of purity and autonomy in mathematical work and in mathematical concepts developed as mathematical activities became institutionalized, professionalized, and bureaucratized.[7] Generational continuity was re-established.

As merchant and industrial capitalist activities developed and spread, the fetishism of commodities was translated into the fetishism of mathematical truths to a degree only approximated by the Greeks. Under the fragmenting social conditions of capitalist development, the fetishistic transformation of mathematics was (as in the Greek case) accompanied by the development of insular and purist ideologies in mathematics, and more generally in science.

Social and Economic Background. Moneyless, self-sufficient monasteries, manors, and feudal estates dominated the social landscape of medieval Europe. Services and payments in kind prevailed in business transactions.[8] The transition from feudalism to capitalism was a prolonged one, and by no means a uniform one.[9] The range of commodities in pre-modern Europe was narrow, and the market and monetary transactions began to dominate Western economy only in the eighteenth century.[10]

The economic parallels between seventh century BC Ionia and thirteenth century Europe are noteworthy. After Venice defeated Constantinople in 1204, a world currency began to emerge. Trade was revived between the Italian cities and the Levant. Florence issued the golden florin in 1252, Genoa the ducat in 1284, and England gold pennies in 1257. By 1350 every major sovereign in Western Europe had issued some form of coin.[11] The slow but steady development of European mathematics can be traced from this period onward, in much the same way that the development of Greek mathematics can be traced from the Ionian commercial revolution. The Thales of this period is Leonardo of Pisa (Fibonacci). He was a merchant, and his mathematical studies mark the dawn of European mathematics. His *Liber Abaci* (1202) reports information on arithmetic and algebra collected during his travels to the Orient. This book was one of the means by which the Hindu-Arabic numeration system was transmitted to Western Europe. Leonardo's discoveries in geometry and trignometry were published in *Practica Geometriae* (1220). While some of his work has no Arabic-Islamic parallels, he was clearly influenced by Arabic-Islamic sources. For example, he quotes Al-Khawarizmi in his discussion of the equation $x^2 + 10x = 39$, and elsewhere.[12]

The Development of the Calculus. The history of European mathematics begins in the twelfth century. It is dominated by the development of the calculus.

Contacts between the Christian and Moslem worlds during the period 1150 to 1250 led to the introduction of Greek, Hindu, and Arabic-Islamic works into Europe. Latin translations of Greek manuscripts were soon forthcoming. Europeans were introduced to Aristotle's works by Jewish and Arab scholars. These works became the foundation upon which high scholasticism was constructed. Discussions of the infinite, infinitesimals, and continuity arose in the context of the progressive revelations of the Greek legacy. The rigor and rationality characteristic of these discussions was stimulated in particular by the combined effects of the economic successes and requirements of twelfth century society, the Aristotelian corpus, and the regulative life of the monasteries. But diversity of opinion was facilitated by the socio-economic diversity stimulated by commercial expansion and demographic change. For example, arguments for the categorematic infinity (the actuality of infinity) were opposed to arguments for the syncategorematic infinity (infinity as a potentiality). These arguments were later drawn on in the systematic development of the calculus.

Between 1350 and 1450 there were large scale outbreaks of the plague and disruptions of the feudal social structure. Mathematical advance was minimal; the two important mathematicians from this period mentioned by D. E. Smith are Oresme and von Gmunden. Already, however, the scholastics had become involved with problems of change and variation. Increasingly, the problems of everyday life focused attention on rates of change. The medieval religious quest which prompted discussion of infinities was at the same time an economic quest – a search for solutions to problems of production and consumption. As commerce and industry developed, the economic motives behind the work on infinities became increasingly explicit.

The rigidity of scholasticism could not survive the crises of the late Middle Ages. Mysticism became an important intellectual strategy; Aristotelianism gave ground to Pythagoreanism and Platonism. This was a manifestation of the intellectual need to break through established and constraining rules of thought which were preventing the solution of new social and economic problems. In particular, problems of change and variation assumed increasing importance as commercial and industrial activities developed, grew in scale, and spread. The Europeans were now ready to move across the threshold reached by the Alexandrians. Their mathematics necessarily began to develop in the direction of the calculus, the mathematics of change and variation.

The period of 1450 to 1650 is one of renewed economic and population growth; it culminates in the first bourgeois revolution, the English Revolution of 1642. Contemporary mathematics developed out of the work on various types of engineering problems carried out by DaVinci, Tartaglia, Cardano, and others. By the late Renaissance, the study of natural science was being transferred from the cloisters to the bourgeois world. This is evident in the work of Stevin and Galileo. The planning of fortifications became such an important context for mathematical work that geometry, surveying, and military architecture were commonly linked by writers on these matters well into the eighteenth century. Napier, Descartes, and Fermat are the three great mathematicians of this era. In the history of the calculus, it is a period of anticipation, and the contributions of Commandino and Kepler stand out.[13] It is at this time that the works of Archimedes begin to influence mathematical developments in Europe.

The Archimedean manuscripts "of the best class" all originated in a manuscript that is no longer extant.[14] The manuscript was in the hands of George Valla, who taught at Venice from 1486 to 1499. Parts of Archimedes' writings were translated in Vallas' book, *De expendedis et fugiendis* (Venice, 1501). The manuscript passed through various hands after Vallas' death in 1499. The three most important manuscripts all contain the same works, including the *quadratura parabolae*. Thus, Archimedes' thoughts on mathematical ideas relevant to the development of the calculus were accessible. During the fifteenth and sixteenth centuries, the manuscripts were in the possession of Regiomontanus (J. Muller), Tartaglia (who translated certain parts into Latin), and Commandino.

Valerio and Stevin were acquainted with Archimedes' works. Kepler, Cavalieri, Roberval, Fermat, G. de Saint-Vincent, Torricelli, and Pascal all re-invented the analytic methods of Archimedes; these methods were not explicitly exhibited in Archimedes' works. Saint-Vincent developed a summation procedure related to but independent of Cavalieri's method of indivisibles (Book VII, *Opus Geometricum*, 1647). He was in possession of Commandino's edition of Archimedes (1558). Cavalieri systematized the method of indivisibles which had shown up periodically in the history of ideas (e.g., Savasorda, the medieval scholastics, DaVinci, Kepler, Galileo). Torricelli believed that the Ancients knew this method but failed to use it in their demonstrations.

From 1650 to 1750 Europe regressed economically. But the stimulus of earlier economic growth had raised problems which were now being pursued by students of the calculus. Military problems were also important stimuli for those people working on the calculus. The works of Barrow, Newton, Leibniz, Huygens, Hudde, and the Bernoullis helped set the mathematical stage for the unfettered economic expansion that began in 1750. The calculus remained in a state of flux between 1750 and 1800, while bourgeois interests were consolidated in the American and French revolutions and the "industrial revolution" got under way. From 1800 on, specialization and professionalization in mathematics increased rapidly. The rigorous formulation of the foundations of the calculus by Cauchy, Dedekind, and Weierstrass followed upon these social developments within the mathematical community.

The Greeks were unable to pursue the problem of motion long enough to anticipate the discoveries of capitalist Europe. Furthermore, the scale of exploration and technology was not great enough to pose insurmountable difficulties in commerce and industry. The problem of motion was immediately taken up when European commerce revived. By 1676, Newton (in his *Quadrature of Curves*) was writing about mathematical quantities "described by a continual motion" and relating this to "the nature of things . . . daily seen in the motion of bodies". The concept of *function* was derived from the study of motion; it was central to practically all mathematical work in the seventeenth and eighteenth centuries, and intimately linked to the calculus.

Capitalism and Calculus. Hessen, as I noted earlier, identified numerous technical problems posed by the developing economy of seventeenth-century Europe, and argued that there was "a complete coincidence of the physical thematics of the period, which arose out of the needs of economics and technique, with the main contents of [Newton's] 'Principia', which in the full sense of the work is a survey and systematic resolution of all the main group of physical problems". He went on to note that while "the economic position is the foundation", scientific theories and work are affected by various superstructures such as "political forms of class war" which are reflected in the minds of participants. To this must be added the effects of the institutionalization of scientific activity, the crystallization of the scientific role, and the creation of more or less stable networks of transportation, communication, and exchange in science. In conformity with Hessen's thesis, the *calculus* developed as a more or less direct response to the major scientific problems of the seventeenth century in Europe; these were in turn rooted in problems raised by the development of merchant capitalism.

There were four major problem domains which guided the development of mathematics in the direction of the calculus. First, in the study of motion in which velocities and accelerations vary from instant to instant, the following problem arose: find the velocity and acceleration of a body at any instant, given the formula for the distance covered by a body as a function of time; and solve the converse problem. Second, the study of optics required that ways be developed to find the tangent to a curve. In order, for example, to apply the law of refraction, it is necessary to know the angle at which a light ray strikes a lens. The problem requires finding the normal or the tangent. Furthermore, the direction of motion of a moving body at any point of its path is the direction of the tangent to the path. Fermat, Descartes, Huygens, and Newton were all directly involved in the design of lenses.

The third type of problem involves finding the maximum or minimum value of a function. For example: find the angle at which a cannon should be inclined to the ground in order to maximize the cannonball's range, or find the maximum heights reached by projectiles fired at various angles. Both of these problems were addressed by Galileo. The study of planetary motions also leads to maxima and minima problems; for example: find the greatest distance and the shortest distance of a planet from the sun.

The fourth type of problem involves finding the lengths of curves (for example, the distance covered by a planet in a given period of time), the areas bounded by curves, volumes bounded by surfaces, centers of gravity, and the gravitational attraction that an *extended* body such as a planet exerts on another body. The theory of infinitesimals dealt exclusively with computations of areas, volumes, and centers of gravity up to the middle of the seventeenth century. The relationship of this work to human productive activity is indicated by Dantzig:[15]

> The importance of infinite processes for the practical exigencies of technical life can hardly be overemphasized. Practically all applications of arithmetic to geometry, mechanics, physics, and even statistics involve these processes directly or indirectly. Indirectly because of the generous use these sciences make of irrationals and transcendentals, directly because of the most fundamental concepts used in these processes.

The notion of length of an arc of a curve, areas, volumes, masses, moments, pressures, forces, stresses, strains, velocities, accelerations, and so on are all features of the rational linear world. Infinitesimals become increasingly central to mathematics as people encounter curvilinear and non-uniform phenomena. The Greeks had begun to encounter the more difficult problems posed by such phenomena, but the decline of their culture aborted the process.

Newton and Leibniz reduced the four problem domains of rates, tangents, maxima and minima, and summations to differentiation and anti-differentiation. The argument in some historical accounts of science that Newton's work reflects a physical orientation whereas Leibniz's work reflects a philosophical orientation should not obscure the fact that Leibniz too had a "physical" orientation. He made contributions to mechanics, optics, hydrostatics, pneumatics, nautical science, and calculating technology.

And his recommendation on the founding of a German academy of science in 1669 (the Berlin Academy was founded in 1700) stressed utility; in opposition to the "monkishness" of the universities, he envisioned a society which would promote inventions in mechanics and chemical and physiological discoveries.

The study of ordinary differential equations was motivated by problems in elasticity (e.g., what shapes are assumed by beams when loads are applied), the pendulum, and astronomy. The approximate period of a pendulum was used to measure the force of gravity at various points of the surface of the earth because the period depends on the acceleration "*g*" determined by that force; pendulum problems were related to studies of the earth's shape, and the verification of the inverse square law. In astronomy, the most important problem was describing the motion of the moon. The determination of a ship's longitude depended on knowing the direction of the moon from a standard position (Greenwich, England from the late eighteenth century onward).

First order ordinary differential equations arose in work on trajectories. John Bernoulli, for example, solved the problem of finding the motion of a projectile in a medium whose resistance is proportional to any power of the velocity. Second order differential equations arose in physical problems as early as 1691. James Bernoulli solved the problem of the shape of a sail under the pressure of the wind. In 1728, Euler considered such equations while working for the King of Prussia on the effect of air resistance on projectiles.

The people who created the calculus worked on the practical problems encountered by merchant and industrial capitalists; solutions to these problems were often facilitated by state patronage. The most mundane considerations were wedded with, not separated from, aesthetics and "pure curiosity", often quite explicitly. Galileo, for example, writing on the cycloid, links aesthetics and bridges:[16]

> More than 50 years ago this curved line came to my mind and I wanted to describe it, admiring it because of its most gracious curvature, adaptable to the arches of a bridge.

Barrow's collection of theorems, *Lectiones geometricae* (London, 1670), deals mostly with tangents, areas, and lengths of arcs; the starting point in these lectures is *the problem of motion*. Huygen's *Horologium oscillatorium* (Paris, 1673), contains important discoveries in mechanics and mathematics; it is a description of the pendulum clock he invented. Huygens was involved in the search for reliable clocks prompted by the needs of navigators, geographers, and astronomers. Leibniz' "methodus de maximus" (1683) was first presented in an article on the laws of refraction and reflection. The *algorithmic imperative* which dominated the formative period in the history of the calculus is manifest in Leibniz' comments on the integral calculus which he first introduced in "De geometria recondita et analysi indivisibilium atque infinitorum" in 1686:

> From it flow all the admirable theorems and problems of this kind with such ease that there is no more need to teach and retain them than for him who knows our present algebra to memorize many theorems of ordinary geometry.

Ball is quite explicit on this algorithmic imperative. In his discussion of the invention of the differential calculus, he writes that "a sort of machine was constructed":[17]

> ... to solve a problem, it was only necessary to put in the particular function dealt with, or the equation of the particular curve or surface considered, and on performing certain simple operations the result came out. The validity of the process was proved once and for all, and it was no longer requisite to invent some special method for every separate function, curve, surface.

And so as the industrial "machine" of the emerging capitalist era was fashioned, so was the "machine" of the calculus fashioned to provide the mathematical fuel for the economic engine.

Virtually all the individuals associated with the development of the calculus prior to the late nineteenth century had a variety of substantial political and/or military experiences (e.g., Pacioli, Pascal, Rolle, Parent, Taylor, Robins, Euler, Lagrange, the Bernoullis, Carnot, Cauchy, Weierstrass, Cantor). Those individuals with little or no political and/or military experiences who worked on the calculus tend to be either associated with religious institutions or among the increasingly numerous "professionals" who reflected the institutionalization of mathematical activities (e.g., Cavalieri, Saint-Vincent). From the nineteenth century on, the "purity" of developments in the calculus becomes quite pronounced and mathematics in general becomes at once more diversified and more specialized. Schools (e.g., logistic, intuitionist, formalist) emerge for the first time (not counting the philosophical-mathematical schools of ancient times, or the relatively loose networks of Cartesians, Newtonians, etc.). There is a tendency for the family backgrounds of the mathematicians associated with the development of the calculus to shift from ministerial during the "period of anticipation" (prior to 1600) to commercial during the period of "rigorous formulation" (after 1800).

As with the case of the Greek philosopher-mathematicians, the Europeans who worked on the calculus were not necessarily *directly* involved in basic sustenance activities (although many were, especially in the early history of the calculus). The critical factor was the relationship between hand and brain, and how this was controlled by mercantile and state interests.

The algorithmic imperative in the development of the calculus, rooted in the interests of the bourgeoisie, was more general. This fact underscores the extent to which mathematics was influenced by economic activity in the period I am considering. Descartes' analytic geometry, the other great contribution in pre-modern European mathematics, was also created under an algorithmic imperative (and was rooted in the Archimedean corpus). It was, in spite of the conflicts between Cartesians and Newtonians, from the very beginning in constant association with the development of the Newtonian-Leibnizian calculus.[18] Just as Ball compares the calculus to a machine, so Boutroux compares Descartes' analytic geometry to modern industrial processes, and writes that with it mathematical research "est devenue à la lettre, un travail de manufacture".

Summary. The general perspective I have sketched in this chapter parallels the perspective sketched in the preceding chapter on Greek mathematics. The calculus was created on the foundations of medieval mathematics and the Greek legacy. The development of the calculus begins essentially where the development of the Greek method of exhaustion ends. The European exploitation of the Greek legacy was possible because of favorable social, economic, and political developments which influenced mathematical work. As in the Greek case, political and economic diversity and decentralization, specialization, and generational continuity were important factors in stimulating innovative mathematical activity. The consolidation of political and economic power in European society, especially from the nineteenth century on has had an impact on mathematics that echoes the impact of Platonist oligarchies on Greek mathematics. The parallel, however, is weak because capitalist imperialism has sustained economic and political expansion without undermining the exchange imperative; and because the complexity of the modern world system has accommodated the intense separation of hand and brain within some segments of the mathematical community, the equally intense separation of mathematical work from everyday life, and the subordination of mathematical work, "applied" and "pure", to political, economic, and military interests.

A Note on Mathematics in Japan and China. The ideas I have explored in this chapter need be sharpened and tested in future studies of mathematics and civilization. In addition to a fuller treatment of the Greek and European cases, analyses of other cases should be carried out. Japan and China are crucial comparative cases. I want to briefly indicate the relevance of these two cases to the problems I have addressed here.

The Japanese developed a "native calculus" in the seventeenth century. In the midst of a general intellectual renaissance, Japanese mathematicians worked on the integral calculus. Contributors to this development include Kittoku (*Ketdugi-sho,* 1660), Teicho (*Dokai-sho,* 1664), and Kazuyuki (*Kokon Sampo-ki*, 1670). The work on a native calculus was preceded by a period in which new cultural contacts were established. The Portuguese arrived in Japan in the middle of the sixteenth century; shortly thereafter, Christian missionaries arrived. The Dutch arrived in 1609. The Portuguese stimulated Japanese trade in two ways; one was by opening up trade with the West; the other was by serving as "middle-men" in the unofficial trade relations between China and Japan. Official relations between China and Japan had been suspended on account of incursions into Chinese territory by Japanese pirates. The seventeenth century witnessed the development of a monetary economy and a commercial revolution.[19]

Toward the end of the sixteenth century, the samurai victories in the clannish military struggles and the leadership of Hideyoshi consolidated military power. The Tokugawa shoguns completed the consolidation by outlawing seafaring and shipbuilding (1626–1628). The internal wars of the military clans and the external ventures of pirates were ended. Under these more peaceful conditions, the merchant class recovered some of the political and economic power lost earlier to the samurai. Financiers and merchants supported the development of a middle-class urban culture in opposition to the samurai. Thus, although the Tokugawa shoguns had consolidated political and military power, there was still some degree of decentralization manifested in the autonomy of the merchant class. The Japanese calculus thus developed in an era of new cultural contacts, and a relative increase in political and economic diversity and decentralization.

The precise influence of Western mathematics on seventeenth-century Japanese mathematics is not known. However, European treatises did find their way into Japan; they were brought in, for example, by the Dutch traders who held a bartering monopoly at the port of Nagasaki. Some of these works may have introduced the Japanese to Western mathematics.

Some Japanese studied and travelled in seventeenth-century Europe. Whatever the precise role of diffusion, this case illustrates the affinity between commercial expansion based on money and exchange, and the calculus.

The Japanese renaissance was short-lived, and the work on the calculus did not lead to a mathematical revolution. The further consolidation of Tokugawa power and the isolationist policies of the shoguns eventually closed Japan to external relations and eliminated the privileges of the merchant class. This also put an end to innovative mathematical work.

It will be useful to review some points about mathematics in China which I discussed in Chapter 11. The Chinese *did not* develop anything to compare with Euclid's *Elements*, nor did they develop a native calculus on the same level as the Japanese.[20] Correspondingly, they did not experience the development of a monetary and exchange economy to the same degree as the Japanese. The history of mathematics and economics in China reflects the fact that ecological and sociocultural conditions in China were not favorable to the development of a diversified economic base and exchange. The functional differentiation of more or less autonomous social activities could not develop to any great extent under such circumstances; and this is reflected in the relative dependence of merchants and mathematicians (along with other social groups) on the state.

The major works in Chinese mathematics were produced over a period of 1500 years, culminating in the "Golden Age" of mathematics in the thirteenth century. These works deal exclusively with practical problems. When Chinese mathematicians did have a chance to specialize, it was in isolation or in subordination to the state. Under these conditions, the generalization of mathematical results meant the generalization of *arithmetic*. This process was never carried very far, but work along these lines led to important contributions in algebra. The Chinese could not develop a "Greek mathematics" because (a) general social, economic, and political conditions prevented the functional differentiation of mathematics, specialization, and generational continuity, and (b) they were therefore unable to exploit the benefits of schools of mathematics and networks of teachers and students with respect to "pure" and "abstract" mathematics. The restriction to arithmetic, incidentally, was not necessarily an impediment to the development of mathematical theory (the Greek view of arithmetic and geometry notwithstanding). Given the opportunity, the Chinese mathematicians might very well have developed a general mathematics more intimately linked to sustenance activities and less static then Greek mathematics.

Chinese mathematicians, employed as provincial civil servants, or in the imperial bureaucracy, were unable to develop the autonomy necessary for establishing schools, specializing within mathematical communities, and theorizing. Not even the thirteenth century Chinese renaissance could offset the inhibiting effects of the Chinese ecological and sociocultural environments.

CONCLUSION

The development of "Greek mathematics" and of the calculus reflects the impact of social, economic, and political factors on mathematics. When, and to the extent that, mathematics becomes a functionally differentiated, institutionally autonomous activity, it begins to "generate" mathematics. In other words, the vulgar notion that "mathematics causes mathematics" (pure mathematics) is based on the fact that in a mathematical "community", mathematicians use the works of their colleagues and of earlier generations as the material resources for their mathematical labors. Systematization, rationalization, generalization, and abstraction in mathematics (as in other human activities) are dependent on organizing mathematical workers in a particular way. These processes can occur only to the extent that (a) mathematics *is* functionally differentiated from other social activities, and (b) there is continuous communication across generations of specialists.

Historically, the functional differentiation of mathematical activity has been associated with the development of exchange-based economies and decentralized political units. To the extent that exchange becomes an end-in-itself the functional differentiation of mathematics will tend to intensify and, as a consequence, mathematics will become increasingly insulated from other social activities. This will also occur where functional differentiation persists in the face of political centralization and loss of the economic diversity on which exchange rests. Thus, mathematical abstraction can be carried out in different mathematical and sociocultural settings. The type of setting determines the type and degree of abstraction, and in particular the extent to which it will be alienated and alienating.

The development of the central themes (especially the calculus and analytic geometry) in Western mathematics reflects the capitalist tendencies of exchange economies. *To the extent* that contemporary mathematics is a product of capitalist culture it is fair to refer to it as "bourgeois mathematics". The most abstract aspects of mathematics cannot escape this designation since they are constructed, level after level, on the social, economic, and political realities of capitalism. But the situation is even more complex.

The complexity of contemporary mathematics reflects the fact that it is rooted in a wide range of experiences that cut across different cultures and different mathematical subcultures. In this exploratory analysis, I have not been able to take into account the range of relationships within mathematics and between mathematical and other social activities. Nor have I been able to carry my discussion toward a more *constructivist* level. Contemporary mathematics is more than "bourgeois mathematics". But even so, it is not "universal" or "eternal"; it remains inextricably tied to the range of activities which human beings have engaged in up to the present time. If mathematics has a capitalist bias today, it is because capitalist society has, more than any other society before it, supported the development of self-aware mathematicians working in an institutionally autonomous social sphere. As we probe more deeply into the micro- and macro-realms of the universe, we extend the range of our experiences. To the extent that these realms differ substantially from our everyday world, they will stimulate new developments in mathematics. At the same time, as we change the social, economic, and political conditions of our lives, we also change the material bases and social structure of mathematics. A radical change in the nature of our social relationships will be reflected in radical changes in how we organize to do mathematics – and these changes will in turn affect how we think about and the content of our mathematics.

NOTES

PART I

CHAPTER 1 & APPENDIX A

[1] J. Needleman, *A Sense of the Cosmos: The Encounter of Modern Science and Ancient Truth*, Doubleday and Company, Garden City, N.Y., 1975; D. Veda, *Zen and Science*, Risosha, Ltd., Tokyo, 1963; K. N. Jayatilleke, *The Message of the Buddha*, The Free Press, New York, 1974, pp. 90–103; G. M. Globus and I. Savodnik (eds.), *Consciousness and the Brain*, Plenum Press, New York, 1976; J. Hayward, 'Toward a Meeting Point of Physics and Buddhism', in R. Fields (ed.), *Loka: A Journal From the Naropa Institute*, Anchor Books, Garden City, N.Y., 1975, pp. 38–39; cf. C. E. M. Joad, *Philosophical Aspects of Modern Science*, George Allen and Unwin, London, 1932, Chapter XI, and especially the remarks on p. 318.

[2] These claims are made, respectively, by Z. D. Sung, *The Symbol Of Yi King*, The China Modern Education Company, Shanghai, 1934; and G. Beau, *La médecine chinoise*, Le Rayon de la Science, Paris, 1965.

[3] The argument in Beau, *op. cit.*, combines all three errors; his "announcement" of Einstein's ancient Chinese anticipator is characterized by Sivin as "a misinterpretation of a mistranslation of a ludicrously misdated text": N. Sivin, 'Preface', N. Sivin and S. Nakayama (eds.), *Chinese Science*, MIT Press, Cambridge, Mass., p. xv.

[4] W. I. Thompson, *Passages About Earth*, Harper and Row, New York, 1974, pp. 187–193.

[5] J. R. Oppenheimer, *Science and the Common Understanding*, Oxford University Press, New York, 1954, pp. 8–9; N. Bohr, *Atomic Physics and Human Knowledge*, John Wiley, New York, 1958, p. 20; W. Heisenberg, *Physics and Philosophy*, Harper Torchbooks, New York, 1958, p. 202; C. H. Townes, 'The Convergence of Science and Religion', *Zygon*, Vol. 1, No. 3 (September), 1966, pp. 301–311.

[6] J. Vigier, 'Possible Internal Subquantum Motions of Elementary Particles', in W. Yourgrau and A. D. Brick (eds.), *Physics, Logic and History*, Plenum Press, New York, 1970, pp. 191–202; E. Wigner, *Symmetries and Reflections*, Indiana University Press, Bloomington, 1967, p. 186, discusses the relevance of consciousness to atomic physics; W. Pauli and C. Jung, *Naturer Klarung und psyche*, Studien aus dem C. G. Jung-Institut, Vol. IV (Zürich), 1952.

[7] I. Barbour, *Issues in Science and Religion*, Harper Torchbooks, New York, 1966; on the seventeenth-century "virtuosi", see R. Westfall, *Science and Religion in Seventeenth-Century England*, University of Michigan Press, Ann Arbor, 1973; representatives of neo-orthodox theology include K. Barth, *Dogmatics in Outline*, SCM Press, London, 1949; W. A. Whitehouse, *Christian Faith and the Scientific Attitude*, Philosophical Library, New York, 1952; and A. van der Zeil, *The Natural Sciences and the Christian Message*, Denison, Minneapolis, 1960. For an existentialist perspective, see M. H. Hartshorne, *The Promise of Science and the Power of Faith*, Westminster Press, Philadelphia, 1958; T. R. Miles' *Religion and the Scientific Outlook*, George Allen and Unwin, London,

1959, is an example of linguistic analysis; on liberal theology, see H. DeWolf, *The Case for Theology in Liberal Perspective*, Westminster Press, Philadelphia, 1959. On natural theology, see A. Compton, *The Human Meaning of Science*, University of North Carolina Press, Chapel Hill, 1970; and A. Eddington, *The Nature of the Physical World*, Cambridge University Press, Cambridge, 1928, and *Science and the Unseen World*, George Allen and Unwin, 1929. A. N. Whitehead's views on process philosophy are presented in *Process and Reality*, Macmillan, New York, 1929.

[8] R. G. H. Siu argues for a supplementary relationship between Taoism and science in *The Tao of Science*, MIT Press, Cambridge, Mass., 1957; on the convergence between science and religion see Townes, *op. cit*. The argument that the world views of physicists and mystics are indistinguishable is defended in L. LeShan, *The Medium, the Mystic, and the Physicist*, Ballantine Books, New York, 1975. I focus my attention on F. Capra's *The Tao of Physics; An Exploration of the Parallels Between Modern Physics and Eastern Mysticism;* Shambhala Publications, Berkeley, Calif., 1975. But I will also refer to G. Zukav, *The Dancing Wu Li Masters*, William Morrow, New York, 1979; Zukav defends the thesis that there are parallels between modern physics and Eastern mysticism in a relatively incidental way compared to Capra. Two recent paperbacks are weak imitations of Capra; Michael Talbot's *Mysticism and the New Physics*, Bantam Books, New York, 1981; and Amaury de Riencourt's *The Eye of Shiva: Eastern Mysticism and Science*, William Morrow, New York, 1981. It is curious that de Riencourt fails to cite not only Capra but Zukav, whose book was, like his own, published as a Morrow Quill Paperback. Talbot's book illustrates the rhetoric of awe in parallelism regarding the results of modern physics and mystical experiences. It also underscores the general parallelist propensity for assuming that of all the scientists, it is the *physicists* (and especially the high energy and relativity theorists) who have hit upon a "truth" that reflects and applies to *all* of reality. De Riencourt's approach illustrates the parallelist tendency to focus on consciousness and to ignore social forces. Knowledge is rooted in consciousness, not in social reality; so in a sort of inverted Durkheimian fashion, knowledge is projected onto a "mystery plane" (the realm of religion) rather than rooted in social relationships. See also K. Wilber (ed.), *The Holographic Paradigm and Other Paradoxes*, Shambhala, Boulder, 1982; and B. Holbrook, *The Stone Monkey, An Alternative Chinese – Scientific Reality*, W. Morrow, New York, 1981.

[9] Capra, *op. cit.*, 160; cf. C. Hooker, 'The Nature of Quantum Mechanical Reality: Einstein vs. Bohr', in R. Colodny (ed.), *Paradigms and Paradoxes*, University of Pittsburgh Press, Pittsburgh, 1972, pp. 206–207; and see G. Holton, *Thematic Origins of Scientific Thought*, Harvard University Press, Cambridge, Mass., 1973, p. 121.

[10] De Riencourt, *op. cit.*, p. 112, argues that Eastern mysticism is more compatible with modern physics than Western mysticism because it is less an art based on love than an experimental science based on a form of "translogical cognition".

[11] E. Conze, *Thirty Years of Buddhist Studies*, University of South Carolina Press, Columbia, 1968, pp. 213–215.

[12] C. R. Otto, *Mysticism East and West*, Macmillan, New York, 1932, p. 5; N. Smart, *Reason and Faith*, Routledge and Kegan Paul, London, 1958, *World Religions: A Dialogue*, Penguin Books, Baltimore, 1969, and *The Concept of Worship*, Macmillan, New York, 1972; see also I. Barbour, *Myths, Models, and Paradigms*, Harper and Row, New York, pp. 78–79.

[13] K. W. Ford, *The World of Elementary Particles*, Blaisdelt, New York, 1965, p. 209.

[14] Walter Sorrell refers to Leonardo's statement as "the stunning aphorism" in his *The Dancer's Image*, Columbia University Press, New York, 1971, p. 254; cf. P. Taylor (ed.), *The Notebooks of Leonardo da Vinci*, The New American Library, 1960, p. 196.
[15] W. Thirring, 'Urbausteine der Materie', *Almanach der Osterreichischen Akademie der Wissenschaften*, Vol. 118 (Vienna), 1968, p. 159; Chang Tsai is quoted in J. Needham, *Science and Civilization in China*, Vol. IV, Part 1, Section 26, Cambridge University Press, New York, 1970, p. 33.
[16] J. Needham, *The Grand Titration*, University of Toronto Press, Toronto, 1969, pp. 250–251; R. G. H. Siu, *Ch'i: A Neo-Taoist Approach to Life*, MIT Press, Cambridge, Mass., 1974, pp. 256–257, and 17–18, 265–266.
[17] G. Chew, *S-Matrix Theory of Strong Interactions*, W. A. Benjamin, New York, 1962, *Strong Interaction Physics*, W. A. Benjamin, New York, 1964, *The Analytic S-Matrix: A Basis for Nuclear Democracy*, W. A. Benjamin, New York, 1966, 'Bootstrap: A Scientific Idea?', *Science*, Vol. **161** (23 May), 1968, pp. 762–765, and 'Hadron Bootstrap: Triumph or Frustration?', *Physics Today*, Vol. **23** (October), 1970, pp. 23–28.
[18] Chew, 1966, *op. cit*., p. 4.
[19] See, for example, F. J. Dyson, 'Mathematics in the Physical Sciences', in Committee on Support of Research in the Mathematical Sciences (eds.), *The Mathematical Sciences*, MIT Press, Cambridge, Mass., 1969; Chew, 1966, *op. cit*., p. 99 characterizes these critics as "uninitiated".
[20] Chew, 1962, *op. cit*., pp. 1–3.
[21] Chew, 1964, *op. cit*., pp. 105–106.
[22] R. Feynman, *The Character of Physical Law*, MIT Press, Cambridge, Mass., 1967, p. 167; and see G. Feinberg, 'The Philosophical Implications of Contemporary Physics', in Colodny, *op. cit*., pp. 33–46.
[23] Conze, *op. cit*.; D. T. Suzuki, *On Indian Mahayana Buddhism*, Harper and Row, New York, 1968; G. C. C. Chang, *The Buddhist Teaching of Totality*, Pennsylvania State University Press, University Park, 1971.
[24] Conze, *op. cit*., p. 213.
[25] Capra, *op. cit*., pp. 295–296; Chang, *op. cit*., p. 255.
[26] Chew, 1968, *op. cit*., p. 273; Wigner, *op. cit*., p. 172.
[27] See the discussion of Bohm's "holonomy physics" in Section III; and see Chapter 2 for his views on lens versus hologram.

CHAPTER 2 & APPENDIX B

[1] Capra, *op. cit*., p. 12.
[2] See the Appendix to this chapter.
[3] See, for example, E. B. Cowell, F. M. Muller, and J. Takakusu (eds.), *Buddhist Mahayana Texts*, Dover, New York, 1969, pp. vi–x (an unabridged and unaltered reprint of Vol. XLIX in 'The Sacred Book of the East' series published in 1894 by the Clarendon Press at Oxford); and S. Radhakrishnan and C. A. Moore (ed.), *A Sourcebook in Indian Philosophy*, Princeton University Press, Princeton, N.J., 1957, pp. ix–xiv.
[4] Feynman, *op. cit*., pp. 39–41.
[5] See the Appendix to this chapter; and cf. W. V. O. Quine, 'Existence', in Yourgrau and Breck, *op. cit*., pp. 89–103.

[6] See, for example, C. W. Rietdijk, *On Waves, Particles, and Hidden Variables*, Van Gorcum, Assen, Holland, 1971, pp. i–iv; and D. Bohm, 'Quantum Theory as an Indication of a New Order in Physics. Part A. The Development of New Orders as Shown Through the History of Physics', and 'Part B. Implicate and Explicate Order in Physical Law', *Foundations of Physics*, Vol. 1, No. 4 (1971), pp. 359–381, and Vol. 3, No. 2 (1973), pp. 139–68.
[7] G. Scholem, 'Mysticism and Society', *Diogenes*, No. 58 (Summer), 1967, pp. 2–3.
[8] M. Gell-Mann and Y. Ne'eman (eds. and authors), *The Eightfold Way*, W. A. Benjamin, New York, 1964, pp. 9–10, 117; cf. Chew, 1968, *op. cit.*, 764; and see G. Chew, M. Gell-Mann, and A. H. Rosenfeld, 'Strongly Interacting Particles', *Scientific American*, Vol. **220**, No. 2 (February), 1964, pp. 74–93.
[9] J. Graves, *The Conceptual Foundations of Contemporary Relativity Theory*, MIT Press, Cambridge, Mass., 1971, pp. 24–25.
[10] Chang, *op. cit.*, pp. 66, 75; Conze, *op. cit.*, p. 213.
[11] See, for example, Radhakrishnan and Moore, *op. cit.*, p. ix.
[12] Cf. J. Ravetz, *Scientific Knowledge and its Social Problems*, Clarendon Press, Oxford, 1971, p. 175; and W. T. Stace, *Mysticism and Philosophy*, Macmillan, London, 1961, pp. 277–306.
[13] See D. T. Suzuki's comments in B. L. Suzuki, *Mahayana Buddhism*, George Allen and Unwin, London, 1959, p. 33; for Minkowski's remarks, see A. Einstein *et al.*, *The Principle of Relativity*, Dover, New York, 1923, p. 75.
[14] Graves, *op. cit.*, p. 61.
[15] C. O. Evans, 'Attention, Meaning, and Altered States of Consciousness', Philosophy Department, Louisiana State University, Baton Rouge, La., undated mimeographed manuscript; cf. M. Polanyi, *Personal Knowledge*, Harper Torchbooks, New York, 1964, pp. 55–65.
[16] Chang, *op. cit.*, pp. 18–19.
[17] Quoted in M. Jammer, *The Philosophy of Quantum Mechanics*, John Wiley, New York, 1974, pp. 86–87.
[18] C. A. Coulson, *Christianity in an Age of Science*, Oxford University Press, London, 1953; Barbour, 1974, *op. cit.*, pp. 84–91; cf. H. A. Bedau, 'Complementarity and the Relation Between Science and Religion', *Zygon*, Vol. **9**, No. 3, September, 1974, pp. 202–223; D. M. MacKay, ' "Complementarity" in Scientific and Theological Thinking', *Zygon*, Vol. 9, No. 3, September, 1974, pp. 225–244; W. H. Austin, 'Complementarity and Theological Paradox', *Zygon*, Vol. 2, No. 4, December, 1967, pp. 365–381.
[19] Jammer, *op. cit.*, pp. 89–90, 95; see also Bohr, 1929, *op. cit.*; and A. Einstein, 'Reply to Criticisms', in P. A. Schilpp (ed.), *Albert Einstein: Philosopher-Scientist*, Harper and Row, New York, 1959, pp. 663–688.
[20] See, for example, J. L. Park, 'Complementarity Without Paradox: A Physicist's Reply to Professor Austin', *Zygon*, Vol. 2, No. 4 (December), 1967, pp. 382–388.
[21] Jammer, *op. cit.*, pp. 103–104; L. Rosenfeld, 'Foundations of Quantum Theory and Complementarity', *Nature*, Vol. **190** (29 April), 1961, pp. 384–388; cf., L. DeBroglie, *The Revolution in Physics*, Noonday Press, New York, 1953, p. 218.
[22] Austin, *op. cit.*; Park, *op. cit.*
[23] See, for example, Colodny, *op. cit.*; A. Einstein, 'Autobiographical Notes', in Schilpp, *op. cit.*, pp. 3–95; P. Chambadal, *Paradoxes in Physics*, Transworld, London, 1973; Y. Terletskii, *Paradoxes in the Theory of Relativity*, Plenum Press, New York, 1968.

[24] D. Finkelstein, 'The Physics of Logic', in Colodny, *op. cit.*, p. 57; cf. D. Bohm, *Quantum Theory*, Prentice-Hall, New York, 1951, pp. 142, 168.
[25] B. D'Espagnat, *Conceptual Foundations of Quantum Mechanics*, W. A. Benjamin, Menlo Park, Calif., 1971, p. 413; B. Kuznetsov, 'Quantum-Relativistic Retrospection and the History of Classical Physics: Classical Rationalism and Nonclassical Science', in R. McCormach (ed.), *Historical Studies in the Physical Sciences*, Vol. 3, University of Pennsylvania Press, Philadelphia, 1971, p. 117 (on Bohr and Einstein); and Terletskii, *op. cit.*, p. 1; cf. I. Shah, *The Way of the Sufi*, E. P. Dutton, New York, 1970, pp. 26–27.
[26] Chang, *op. cit.*, p. x; cf. Stace, *op. cit.*, pp. 251–271.
[27] Capra, *op. cit.*, pp. 130, 140, 289.
[28] Chang, *op. cit.*, pp. 83, 253, 257, 261.
[29] Bohm, 1951, *op. cit.*, pp. 139–140.
[30] Jammer, *op. cit.*, p. 199; Graves, *op. cit.*, pp. 57ff., 154.
[31] Terletskii, *op. cit.*, pp. 54, 60.
[32] Graves, *op. cit.*, pp. 314, 235–236.
[33] Chang, *op. cit.*, pp. 11–12, 20–21.
[34] Feinberg, *op. cit.*, pp. 33–45; cf. Bohm, 1951, *op. cit.*, p. 168.
[35] Holton, *op. cit.*, pp. 24–26.
[36] G. Gale, 'Chew's Monadology', *Journal of the History of Ideas*, Vol. **35** (April–June), 1974, pp. 339–348.
[37] Graves, *op. cit.*, p. 216; D'Espagnat, *op. cit.*, p. 248; on the question of the conjunction between nonlinearity and consciousness, cf. Hooker, *op. cit.*, p. 308.
[38] Bohm, 1971, 1973, *op. cit.* (esp. pp. 143–147); cf. J. Smuts, *Holism and Evolution*, Greenwood Press, Westport, Conn., 1973 (reprint of the 1926 edition published by Macmillan Ltd., London).
[39] Siu, 1974, *op. cit.*, pp. 69–84.
[40] W. TenHouten and C. D. Kaplan, *Science and its Mirror Image*, Harper and Row, New York, 1973, pp. 194–224.
[41] See the discussion on "split-brain" research in the Appendix to this chapter.
[42] Needleman, *op. cit.*, pp. 104–105; T. Roszak, *Where the Wasteland Ends*, Anchor Books, Garden City, N.Y., 1973, p. 417.
[43] Holton, *op. cit.*, pp. 275–280; C. A. Hooker, 'Philosophy and Metaphilosophy of Science: Empiricism, Popperianism and Realism', *Synthese*, Vol. **32**, 1975, pp. 177–231; cf. C. Tart, 'States of Consciousness and State-Specific Sciences', in R. E. Ornstein (ed.), *The Nature of Human Consciousness*, W. H. Freeman, San Francisco, 1973, pp. 41–60.
[44] R. Ravindra, 'Experiment and Experience: A Critique of Modern Scientific Knowing', *Dalhousie Review*, Vol. 55, 1975–1976, p. 670.
[45] See J. J. C. Smart, *Philosophy and Scientific Realism*, Humanities Press, New York, 1963; cf. Graves, *op. cit.*, p. 7.
[46] D'Espagnat, *op. cit.*, p. 420.
[47] Cf. Sivin, *op. cit.*, pp. xiv–xviii.
[48] Bohm, 1951, *op. cit.*, pp. 169–171.
[49] *Ibid.*, p. 168.
[50] Capra, *op. cit.*, p. 142.
[51] *Ibid.*, p. 25.

[52] H. Robinson, *Renascent Rationalism*, Macmillan, Toronto, 1975, pp. 106–107.
[53] L. L. Whyte, *The Universe of Experience: A World View Beyond Science and Religion*, Harper Torchbooks, New York, 1974, pp. 64, 120, 141–142 (note 5).
[54] T. S. Kuhn, *The Structure of Scientific Revolutions*, 2nd ed., University of Chicago Press, Chicago, 1970.
[55] P. Forman, 'Weimar Culture, Causality and Quantum Theory, 1918–1927: Adaptation by German Physicists and Mathematicians to a Hostile Intellectual Environment', in R. McCormach, *op. cit.*, pp. 1–115; L. Feuer, *Einstein and the Generations of Science*, Basic Books, New York, 1974.
[56] This appendix is excerpted from S. Restivo, 'Joseph Needham and the Comparative Sociology of Chinese and Modern Science', in R. A. Jones and H. Kuklick (eds.), *Research in Sociology of Knowledge, Sciences and Art*, Vol. II, JAI Press, Greenwich, Conn., 1979, pp. 25–51. For specific citations to Needham's work, readers are referred to the original article. References for other authors cited in this excerpt are as follows: Sivin, *op. cit.*, and personal correspondence; W. McNeill, 'Review of J. Needham, The Grand Titration', *Science*, Vol. 67 (23 January), 1970, p. 367; D. Campbell, 'Distinguishing Differences of Perception from Failures of Communication in Cross-Cultural Studies', in F. Northrop and H. Livingston (eds.), *Cross-Cultural Understanding: Epistemology in Anthropology*, Harper and Row, New York, 1964, pp. 308–336; A. Paredes and M. Hepburn, 'The Split Brain and the Culture and Cognition Paradox', *Current Anthropology*, Vol. 17, No. 9 (March), 1976, pp. 121–127.

CHAPTER 3

[1] Capra, *op. cit.*, p. 46.
[2] *Ibid.*, pp. 13, 19.
[3] H. Shih, 'Ch'an (Zen) Buddhism in China, its History and Method', *Philosophy East and West*, Vol. 3, No. 1 (April), 1953, pp. 3–24; E. Benz, 'Buddhism in the Western World', in H. Dumoulin and J. C. Maraldo (eds.), *Buddhism in the Modern World*, Macmillan, New York, 1976, pp. 305–322.
[4] S. Radhakrishnan, *Eastern Religion and Western Thought*, 2nd ed., Oxford University Press, New York, 1940, pp. ix, 61–63.
[5] Roszak, *op. cit.*, pp. 367, 422, 228.
[6] LeShan, *op. cit.*, p. 254; cf. C. D. Broad, *Religion, Philosophy and Psychical Research*, Routledge and Kegan Paul, London, 1953, p. 242; and E. Underhill, *Mysticism*, 4th ed., Methuen, London, 1912, p. 3.
[7] Capra, *op. cit.*, p. 303.
[8] *Ibid.*, p. 85.
[9] *Ibid.*, pp. 12–13.
[10] E. T. Bell, *Mathematics, Queen and Servant of Science*, McGraw-Hill, New York, 1951, p. 21; E. Rignano, *The Psychology of Reasoning*, Harcourt, Brace, New York, 1927, p. 185; A. Woodcock and M. Davis, *Catastrophe Theory*, E. P. Dutton, New York, 1978; O. Spengler, *The Decline of the West*, 2 Vols., A. A. Knopf, New York, 1926 and 1928, Vol. I, pp. 53–89.
[11] Cf. R. Lowie, *Primitive Religion*, Boni and Liveright, New York, 1948; G. Swanson, *The Birth of the Gods*, University of Michigan Press, Ann Arbor, 1960, p. 11; E. B. Tylor, *Primitive Culture*, 3rd ed., Vol. 2, Henry Holt and Co., New York, 1889.

[12] G. Gurvitch, *The Social Frameworks of Knowledge*, Harper Torchbooks, New York, 1971, pp. 37–38; W. Stark, *The Sociology of Knowledge*, Routledge and Kegan Paul, London, 1958, p. 209; M. Scheler, *Die Wissensformen und die Gesellschaft*, Francke Verlag, Bern, 1960, pp. 52–69. Weber's ideas will be sketched below.
[13] Tylor, *op. cit.*, p. 415.
[14] See, for example, the following: E. Norbeck, *Religion in Primitive Society*, Harper and Row, New York, 1961, pp. 86, 228; H. A. Junod, *The Life of a South African Tribe*, 2 Vols., Macmillan, London, 1913; B. A. Warwick, *The Swazi*, Cambridge University Press, Cambridge, Mass., 1940; E. E. Evans-Pritchard, *Nuer Religion*, The Clarendon Press, Oxford, 1956; and M. J. Field, *Religion and Medicine of the Ga People*, Oxford University Press, London, 1937; and see H. Bridges, *American Mysticism*, Harper and Row, New York, 1970; and C. Geertz, 'Religion as a Cultural System', in C. C. Hughes (ed.), *Make Men of Them*, Rand McNally, Chicago, 1972, pp. 221–222.
[15] E. Durkheim, *The Elementary Forms of the Religious Life*, Collier Books, New York, 1961, pp. 465, 468.
[16] *Ibid.*, p. 471.
[17] Underhill, *op. cit*, Chapter 1 and Part II.
[18] Geertz, *op. cit.*, p. 229.
[19] Scholem, *op. cit.*, pp. 2–3.
[20] R. F. Gombrich, 'Ancient Indian Cosmology', in C. Blacken and M. Loewe (eds.), *Ancient Cosmologies*, George Allen and Unwin, London, 1975, pp. 111–112.
[21] L. Dumont, *Homo Hierarchicus*, University of Chicago Press, Chicago, 1970, p. 193.
[22] *Ibid.*, pp. 35, 37, 43–44.
[23] Cf. R. Bendix, *Max Weber: An Intellectual Portrait*, Anchor Books, Garden City, N.Y., 1962, pp. 151–152, 158; M. Weber, *The Religion of India*, The Free Press, Glencoe, Ill., 1958.
[24] C. Eliot, *Hinduism and Buddhism*, 3 Vols., Barnes and Noble, New York, Vol. I, 1954, pp. 41, 48, 166–167 (orig. publ. 1921).
[25] Bendix, *op. cit.*, pp. 155–156.
[26] S. L. Jaki, *Science and Creation*, Scottish Academic Press, Edinburgh, 1974, pp. 5–7; cf. V. G. Aiyer, *The Chronology of Ancient India*, G. A. Nateson, Madras, 1901, pp. 112–116; and H. H. Wilson, *The Vishnu Purana: A System of Hindu Mythology and Tradition*, John Murray, London, 1840.
[27] M. Monier-Williams, *Buddhism*, 2nd ed., John Murray, London, 1890, p. 224.
[28] K. K. S. Ch'en, *Buddhism*, Barron's Educational Series, New York, 1968, pp. 76–77.
[29] P. A. Pardue, *Buddhism*, Macmillan, New York, 1971, pp. 3–5; L. Stryk (ed.), *World of the Buddha*, Anchor Books, Garden City, N.Y., 1969, pp. xxxv–xxxvi; Monier-Williams, *op. cit.*, p. 36; M. Stevenson, *The Heart of Jainism*, Oxford University Press, London, 1915, p. 2.
[30] Pardue, *op. cit.*, pp. 6–7.
[31] *Ibid.*, pp. 8–9.
[32] *Ibid.*, pp. 25–26; Monier-Williams, *op. cit.*, p. 45.
[33] See E. M. Layton's survey, *Buddhism in America*, Nelson-Hall, Chicago, 1976.
[34] M. Spiro, *Buddhism and Society: A Great Tradition and its Burmese Vicissitudes*, Harper and Row, New York, 1970, pp. 9–11.
[35] Spiro follows F. Edgerton, 'Dominant Ideas in the Formation of Indian Culture', *Journal of the American Oriental Society*, Vol. 62, 1942, pp. 151–156 in distinguishing

these two normative systems; see also D. T. Suzuki, *Outlines of Mahayana Buddhism*, Schocken Books, New York, 1963, p. 85 (originally published 1907); cf. Monier-Williams, *op. cit*., p. 4.

36 Weber, *op. cit*., pp. 225–228.

37 Suzuki, 1963, *op. cit*., pp. 4, 56.

38 Monier-Williams, *op. cit*., p. 37.

39 Spiro, *op. cit*., p. 282; T. W. Davids and C. A. F. Rhys Davids, *Dialogues of the Buddha*, Part I, Oxford University Press, New York, 1921, p. 79.

40 Pardue, *op. cit*., pp. 34–55.

41 *Ibid*., p. 51.

42 A. Watts, 'Prefatory Essay', in D. T. Suzuki, 1963, *op. cit*., pp. x, xii; Spiro, *op. cit*., p. 40.

43 Shih, *op. cit*., 3–24.

44 Benz, *op. cit*., pp. 306–307, 317.

45 Layman, *op. cit*., pp. 49–50.

46 On Bruno, see F. Yates, *Giordano Bruno and the Hermetic Tradition*, University of Chicago Press, Chicago, 1964, pp. 331–332; on Don Juan, Castañeda's alleged mentor (the veracity of Cãstaneda's account has been questioned by some critics), see D. Noel (ed.), *Seeing Castañeda*, G. P. Putnam's Sons, New York, 1976, p. 99.

47 S. Dutt, *Buddhist Monks and Monasteries of India*, George Allen and Unwin, London, 1962, pp. 40–41.

48 M. Weber, in H. Gerth and C. W. Mills (eds.), *From Max Weber*, Oxford University Press, New York, 1958, pp. 267–287.

49 J. Freund, *The Sociology of Max Weber*, Vintage, New York, 1969, pp. 177, 197–199.

50 Bendix, *op. cit*., pp. 269–271.

51 Cf. LeShan, *op. cit*., p. 273; W. R. Inge, *Christian Mysticism*, Meridian Books, New York, 1950, p. xvii.

52 Bendix, *op. cit*., pp. 172, 183ff., 198.

53 R. Collins, 'Some Sociological Principles of Heresy, Religious and Secular', paper presented at the Institute for Advanced Studies, Princeton, New Jersey, February, 1975a, and 'Toward a Modern Science of the Occult', *Consciousness and Culture*, Vol. 1, No. 1, 1977, pp. 43–58; cf. P. Bourdieu, 'The Specificity of the Scientific Field and the Social Conditions of the Progress of Reason', *Social Science Information*, Vol. 14, No. 6, 1975, pp. 30–31; and see A. Schutz, *The Problem of Social Reality: Collected Papers*, Vol. I, M. Nijhoff, The Hague, 1976, pp. 207, 229–233; R. Horton, 'African Traditional Thought and Western Science', *Africa*, Vol. 37, 1967, pp. 50–71, 155–187; H. Nowotny, 'Scientific Purity and Nuclear Danger', in E. Mendelsohn, P. Weingart, and R. Whitley (eds.), *The Social Production of Scientific Knowledge*, D. Reidel Publishing Co., Boston, 1977, p. 249.

54 P. Worsley, 'Religion as a Category', in R. Robertson (ed.), *Sociology of Religion*, Penguin Books, Baltimore, 1969, p. 229; J. J. Bachofen, *Das Mutterrecht*, Krais and Hoffman, Stuttgart, 1861; A. D. Malefijit, *Religion and Culture*, Macmillan, New York, 1968, pp. 149–150; W. Goode, *Religion Among the Primitives*, The Free Press, Glencoe, Ill., 1951; and see especially M. Smith, *Jesus the Magician*, Harper and Row, New York, 1978.

55 On language games, see L. Wittgenstein, *Philosophical Investigations*, 3rd ed.,

Macmillan, New York, 1973, and *Tractatus Logico-Philosophicus*, New ed., Humanities Press, Atlantic Highlands, N.J., 1974; E. R. MacCormac, *Metaphor and Myth in Science and Religion*, Duke University Press, Durham, N.C., 1976, pp. 135–136, 139, 143–144.

56 Geertz, *op. cit.*, pp. 224–225; cf. Spengler, *op. cit.*

57 J. M. Yinger, 'Countercultures and Social Change', *American Sociological Review*, Vol. 42, No. 6 (December), 1977, p. 839; Stryk, *op. cit.*, p. xlvii; E. Tiryakian, 'Toward the Sociology of Esoteric Culture', *American Journal of Sociology*, Vol. 79, No. 3 (November), 1972, p. 442.

58 H. V. Guenther, *Tibetan Buddhism Without Mystification*, E. J. Brill, Leiden, 1966, p. vii.

59 Whyte, *op. cit.*, pp. 16, 95–96; cf. H. K. Schilling, *The New Consciousness in Science and Religion*, United Church Press, Philadelphia, 1973, p. 178.

60 C. O. Evans, *The Subject of Consciousness*, Humanities Press, New York, 1970, pp. 100–101, 106, 202.

61 E. Bourguignon, 'Introduction', in E. Bourguignon (ed.), *Religion, Altered States of Consciousness, and Social Change*, Ohio State University Press, Columbus, Ohio, 1973, pp. 5, 9, 23; cf. M. Douglas, *Natural Symbols*, Barrie and Jenkins, London, 1970, p. 81.

62 I. M. Lewis, 'Spirit Possession and Deprivation Cults', *Man*, Vol. I, 1966, p. 318.

63 Bourguignon, *op. cit.*, pp. 29–31.

64 See, for example, F. Staal, *Exploring Mysticism*, University of California Press, Berkeley, 1975, pp. 99–100.

65 See, for example, R. Jones, *Mysticism and Democracy in the English Commonwealth*, Octagon, New York, 1965; cf. J. Wach, *Sociology of Religion*, University of Chicago Press, Chicago, 1944, p. 164.

66 J. Needham, *Science and Civilization in China*, Vol. 2, Cambridge University Press, New York, 1956, pp. 97–98.

67 L. Vesey, *The Communal Experience: Anarchist & Mystical Communities in Twentieth-Century America*, University of Chicago Press, Chicago, 1978, pp. 207–208; cf. S. Cotgrove, 'Styles of Thought', *The British Journal of Sociology*, Vol. **XXXIX**, No. 3 (September), 1978, pp. 364–366.

68 Needham, 1956, *op. cit.* My later remarks on the positive function of a "lack of rigor" for overcoming rigidity in a knowledge system should recall Needham's hypothesis for the reader. It is important to approach the notion of a sociology of mysticism with an appreciation for Bennetta Jules-Rosette's cautions on evaluating non-Western thought forms as maladaptive, and confusing them with altered states of consciousness: see her 'The Politics of Paradigms: Contrasting Theories of Consciousness and Society', paper presented at the meeting of the Humanistic Transdisciplinary Association, Chicago, Ill., January 30, 1977. Other problems and perspectives relevant to the program for a psychosocial theory of mysticism are discussed in the works of Runzo, Richards, and Horne who suggest, respectively, that (a) vision rules in representational art are operative in certain visions, (b) the meaning of *sunyata* should be sought in its use in life rather than located in an "objective referent or counterpart", and (c) there are variations and incommensurabilities in mystical experiences: J. Runzo, 'Visions, Pictures, and Rules', *Religious Studies*, Vol. 13, No. 3 (September), 1977, pp. 303–318; G. Richards, 'Sunyata: Objective Referent or Via Negativa?', *Religious Studies*, Vol. 14, No. 2 (June),

1978, pp. 251–260; J. R. Horne, 'How to Describe Mystical Experiences', *Studies in Religion*, Vol. 6, No. 3, 1976–1977, pp. 279–284. Horne's view of mystical experience as "one of the most disorderly experiences in life", should be compared with Stace's, *op. cit.*, view that mysticism is continous with everyday experience. The controversial problem of the relationship between psychopathology and mysticism must also be considered: see J. White (ed.), *The Highest State of Consciousness*, Anchor Books, Garden City, New York, 1972. For a persuasive argument against the legitimacy (conceptually and empirically) of relating mysticism to regressive, pathological behavior, see R. W. Hood, Jr., 'Psychological Strength and the Report of Intense Religious Experience', *Journal for the Scientific Study of Religion*, Vol. 13, 1974, pp. 65–71; cf. R. C. Zaehner, *Mysticism: Sacred and Profane*, Oxford University Press, London, 1961; and S. Arieti, *The Intrapsychic Self*, Basic Books, New York, 1967, pp. 417–435. I should also mention here the work of T. Ribot, *The Psychology of Attention*, 5th ed., The Open Court, Chicago, 1903, who analyzes St. Teresa's seven stations of prayer or meditation as stages of increasing concentration culminating in a homogeneous state ("absolute monoisleism"); and K. Pribram's view of mystical processes as "unlocalized" patterns of thought, outlined in *Languages of the Brain*, Prentice-Hall, Englewood Cliffs, N.J., 1975.

CHAPTER 4

[1] Capra, *op. cit.*, pp. 12–13.
[2] Zukav, *op. cit.*, pp. 275–276.
[3] G. Feinberg, *What is the World Made Of*?, Anchor Books, Garden City, N.Y., 1978, pp. xii–xiii; Feynman, *op. cit.*, p. 167.
[4] Capra, *op. cit.*, pp. 17–19.
[5] R. Merton, *Social Theory and Social Structure*, enlarged edition, The Free Press, New York, 1968, p. 661; and L. A. Farrall, 'Controversy and Conflict in Science: A Case Study – The English Biometric School and Mendel's Laws', *Social Studies of Science*, Vol. 5, No. 3 (August), 1975, pp. 269–301; and see E. Frankel, 'Corpuscular Optics and the Wave Theory of Light: The Science and Politics of a Revolution in Physics', *Social Studies of Science*, Vol. 6, No. 2 (May), 1976, pp. 141–184; B. Wynne, 'Physics and Psychics: Science, Symbolic Action, and Social Control in Late Victorian England', in B. Barnes and S. Shapin (eds.), *Natural Order*, Sage, London, 1979, pp. 167–186; S. Woolgar, 'Writing an Intellectual History of Scientific Development: The Use of Discovery Accounts', *Social Studies of Science*, Vol. 6, Nos. 3, 4 (September), 1976, pp. 395–422; Forman, *op. cit.*
[6] Forman, *op. cit.*, p. 5.
[7] Forman, *op. cit.*, p. 3, cites Jammer, *op. cit.*; see also pp. 6, 11–13; cf. L. M. Branscomb, 'Taming Technology', *Science*, Vol. 171 (March 12), 1971, pp. 972–977. This exercise in "parallelism" is, of course, subject to my pitfalls arguments.
[8] Forman, *op. cit.*, pp. 16–17.
[9] *Ibid.*, pp. 48–49.
[10] *Ibid.*, pp. 50–51.
[11] *Ibid.*, p. 86.
[12] *Ibid.*, p. 103.

[13] *Ibid.*, p. 166.
[14] B. Harvey, 'Plausibility and the Evaluation of Knowledge: A Case Study of Experimental Quantum Mechanics', *Social Studies of Science*, Vol. 11, No. 1 (February), 1981, pp. 95–130.
[15] H. Collins, 'The Seven Sexes: A Study in the Sociology of a Phenomenon, or the Replication of Experiments in Physics', *Sociology*, Vol. 9, 1975, p. 207.
[16] This discussion is based on K. Knorr, 'The Ethnography of Laboratory Life', in S. Restivo (ed.), *New Directions in the Sociology of Science*, special issue of the International Society for the Sociology of Knowledge Newsletter, Vol. 7, Nos. 1 and 2 (May), 1981, pp. 4–9.
[17] B. Latour and S. Woolgar, *Laboratory Life*, Sage, Beverly Hills, 1979, pp. 51–52, 245.
[18] K. Knorr and D. Knorr, 'From Scenes to Scripts: On the Relationship Between Laboratory Research and Published Paper in Science', *Research Memorandum No. 132*, Institute for Advanced Studies, Vienna (August), 1978; M. Zenzen and S. Restivo, 'The Mysterious Morphology of Immiscible Liquids: A Study of Scientific Practice', *Social Science Information*, Vol. 21, No. 3, 1982, pp. 447–473.
[19] *Ibid.*, p. 448.
[20] L. Fleck, *Genesis and Development of a Scientific Fact*, University of Chicago Press, Chicago, 1979, pp. 95, 100–102 (originally published in German in 1935).
[21] See Spengler's remarks on mathematics in Chapter 2 of Vol. I, *op. cit.*; and pp. 378–379 for the following quotation.

CHAPTER 5

[1] B. Barnes, *Interests and the Growth of Knowledge*, Routledge and Kegan Paul, London, 1977, pp. 10–19, 81; see also S. Shapin, 'Homo Phrenologicus: Anthropological Perspectives on an Historical Problem', in Barnes and Shapin (eds.), *op. cit.*, pp. 43–50; and cf. J. Habermas, *Knowledge and Human Interests*, Beacon Press, Boston, 1971, pp. 196–212, 308–314; and see Barnes and Shapin, *op. cit.*, p. 188. See also S. Woolgar, 'Interests and Explanation in the Social Study of Science.' *Social Studies of Science*, Vol. 11, No. 3 (August), 1981, pp. 365–94.
[2] Capra, *op. cit.*, p. 25.
[3] Zukav, *op. cit.*, pp. 23–25, 54.
[4] Feinberg, *op. cit.*, p. ix.
[5] Capra, *op. cit.*, pp. 66–67; Zukav, *op. cit.*, pp. 88–90.
[6] G. A. Shook, *Mysticism, Science and Revelation*, G. Ronald, Oxford, 1974.
[7] Capra, *op. cit.*, pp. 25, 307; Zukav, *op. cit.*, p. 330.
[8] I. Asimov, 'Scientists and Sages', *The New York Times* (July 27), 1978, p. 19.
[9] I. Asimov, 'Do Scientists Believe in God?', *Gallery*, Vol. 7, No. 7 (June), 1979, pp. 51–52, 104–106.
[10] Cf. W. Schumaker, *The Occult Sciences in the Renaissance*, University of California Press, Berkeley, Calif., 1972, pp. xv, xvi, 7; P. Rossi, 'Hermeticism, Rationality, and the Scientific Revolution', in M. Bonelli and W. Shea (eds.), *Reason, Experiment, and Mysticism in the Scientific Revolution*, Science History Publication, New York, 1975, pp. 271–273.

[11] See, for example, S. H. Nasr, *The Encounter of Man and Nature*, George Allen and Unwin, London, 1968, pp. 51–80, 105–106; Sivin, *op. cit*., pp. xi–xxxvi.
[12] W. Stegmüller, *The Structure and Dynamics of Theories*, Springer-Verlag, New York, 1976, pp. 207–211.
[13] T. Roszak, *The Making of a Counter Culture*, Anchor Books, Garden City, N.Y., 1969, pp. 127–128, 158; cf. Schumaker, *op. cit*., p. 7.
[14] Thompson, *op. cit*., pp. 88, 114–115; G. Krishna, *The Biological Bases of Religion and Genius*, Harper and Row, New York, 1972; C. von Weizsäcker, 'Introduction', in Krishna, *op. cit*., pp. 1–47.
[15] For example, Tiryakian, *op. cit*.; and E. Tiryakian (ed.), *On the Margin of the Visible*, John Wiley, New York, 1974; Schumaker, *op. cit*., p. xvi.
[16] P. Frank, *Modern Science and its Philosophy*, Harvard University Press, Cambridge, Mass., 1941; T. Schnelle and W. Baldamus, 'Mystic Modern Science? Sociological Reflections on the Strange Survival of the Occult Within the Rational Mechanistic World View', *Zeitschrift Für Soziologie*, Vol. 7 (July), 1978, pp. 251–266.
[17] H. Reynolds, 'Mythomystes', in E. T. Taylor (ed.), *Literary Criticism of Seventeenth-Century England*, A. A. Knopf, New York, 1967, pp. 247–248; Yates, *op. cit*., p. 383.
[18] C. Butler, *Number Symbolism*, Barnes and Noble, New York, 1970, p. 23.
[19] Schumaker, *op. cit*., pp. 201, 256; Yates, *op. cit*., p. 426.
[20] *Ibid*., p. 336; and see E. M. Butler, *The Myth of the Magus*, Cambridge University Press, Cambridge, 1948, pp. 51–52, 146.
[21] Yates, *op. cit*., pp. 118, 150–151, 272.
[22] *Ibid*., p. 297.
[23] *Ibid*., p. 264; cf. Jaki, *op. cit*., p. 3.
[24] Yates, *op. cit*., p. 362.
[25] *Ibid*., pp. 358–359.
[26] I. Bentov, *Stalking the Wild Pendulum*, E. P. Dutton, New York, 1977; Yates, *op. cit*., pp. 176, 245, 247, 426.
[27] Ravetz, *op. cit*., pp. 219–223; J. Ravetz, 'Criticisms of Science', in I. Spiegel-Rösing and D. Price (eds.), *Science, Technology, and Society*, Sage, Beverley Hills, 1977, pp. 74–75; H. Kearney, *Science and Change, 1500–1800*, Weidenfield, London, 1971; F. Manuel, *A Portrait of Isaac Newton*, The Clarendon Press, Oxford, 1974, p. 38.
[28] W. Hitchcock, *The Religion of Geology*, Phillips, Campson, & Co., Boston, 1855. Citations for authors mentioned in the concluding paragraphs of this chapter are: P. Duhem, *The Aim and Structure of Physical Theory*, Princeton University Press, Princeton, 1954; G. C. Henry, Jr., *Logos: Mathematics and Christian Theology*, Bucknell University Press, Lewisburg, Penn., 1976; W. O. Greenwood, *Biology and Christian Belief*, Student Christian Movement Press, London, 1938; A. Hardy, *The Biology of God*, Taplinger, New York, 1975; K. F. Mather, *Science in Search of God*, Henry Holt and Co., New York, 1928; Eddington, *op. cit*., and *Space, Time and Gravitation*, Cambridge University Press, Cambridge, 1920; and see L. P. Jacks, *Sir Arthur Eddington: Man of Science and Mystic*, Cambridge University Press, Cambridge, 1949; D. J. Kevles, *The Physicists*, A. A. Knopf, New York, 1977, pp. 177–180, discusses Millikan.
[29] Spiro, *op. cit*., p. 70; and cf. L. Festinger, *A Theory of Cognitive Dissonance*, Row, Peterson, and Co., New York, 1957; on coherence theory, see W. Eckhart, 'Limits to Knowledge', *Limits to Violence Project Paper #10*, Canadian Peace Research Institute, Oakville, Ontario, 1976. For an example of West-West parallelism and *Aristotle* as an

anticipator of twentieth-century physicists, see Chapter 4 in S. Bochner, *The Role of Mathematics in the Rise of Science*, Princeton University Press, Princeton, 1981 (orig. publ. 1966); for a more elaborate and complex but still vulnerable treatment of "homologies" between (a) Pythagorean and Platonic wisdom and (b) modern physics see 'The Structure of Ancient Wisdom – A Symposium', *Journal of Social and Biological Structures*, Vol. 5, No. 3 (July), 1982, pp. 223–288.

CHAPTER 6 & APPENDIX C

[1] J. E. Hofmann, *The History of Mathematics to 1800*, Littlefield, Adams and Co., Totowa, N.J., 1967, pp. 107–108; cf. T. Dantzig, *Number: The Language of Science*, 4th ed., The Free Press, New York, 1954, p. 81; D. Bloor, *Knowledge and Social Imagery*, Routledge and Kegan Paul, London, 1976, pp. 98ff., 107; R. Wilder, *Evolution of Mathematical Concepts*, Halsted Press, New York, 1975, p. 2; J. R. Newman, *Science and Sensibility*, abr. ed., Anchor Books, Garden City, N.Y., 1963, p. 9; and see C. Butler, *op. cit*., pp. 156, 164–165.

[2] Yates, *op. cit*., pp. 151, 449–453.

[3] See, for example, S. Restivo, 'Towards a Sociology of Objectivity', *Sociological Analysis and Theory*, Vol. V, No. 2 (June), 1975, pp. 160–162; J. Ziman, *Public Knowledge*, Cambridge University Press, London, 1968, p. 65.

[4] Kevles, *op. cit*., pp. 400, 403.

[5] For the rhetorical treatment see A. Thackray, 'Reflections on the Decline of Science and Some of its Causes', *Science*, Vol. **173** (July 2), 1971, pp. 27–31; and H. Brooks, 'Can Science Survive in the Modern Age'?, *Science*, Vol. **174** (October 1), 1971, pp. 21–30; cf. K. Boulding, 'What Went Wrong, if Anything, Since Copernicus'?, *Science and Public Affairs*, Vol. **XXX**, No. 1 (January), 1974, p. 23.

[6] Bohm's views are discussed in Appendix C. R. Quinney, 'A Transcendental Way of Knowing', in N. M. Regush (ed.), *Visibles and Invisibles*, Little, Brown and Co., Boston, 1973, p. 168; N. Hawthorne, 'The Artist of the Beautiful', in N. Hawthorne, *Mosses From the Old Manse*, Houghton Mifflin and Co., New York, 1882, p. 213; J. Grange, 'Magic, Technology and Being', *Religious Humanism*, Vol. **III**, No. 2 (Spring), 1974, p. 88.

[7] A. Gouldner, 'The Dark Side of the Dialectic: Toward a New Objectivity', *Sociological Inquiry*, Vol. **46**, No. 1, 1976, p. 7.

[8] R. Horton, *op. cit*.; and see R. Finnegan and R. Horton (eds.), *Modes of Thought*, Faber, London, 1973; P. Wright, 'Astrology and Science in Seventeenth-Century England', *Social Studies of Science*, Vol. **5**, No. 4 (November), 1975, pp. 399–422; cf. M. Foucault, *The Order of Things*, Tavistock, London, 1970, p. 43; and F. Jacob, *La Logigue du Vivant*, Gallinard, Paris, 1970.

[9] M. Douglas, *Purity and Danger*, Routledge and Kegan Paul, London, 1966, pp. 74–93; cf. E. A. Burtt, *The Metaphysical Foundations of Modern Physical Sciences*, Anchor Books, Garden City, N.Y., 1954, pp. 303ff. (orig. published 1924; 2nd ed., 1934); R. Johnston, 'Contextual Knowledge: A Model for the Overthrow of the Internal/External Dichotomy in Science', *Australian and New Zealand Journal of Sociology*, Vol. **12**, No. 3 (October), 1976, p. 199.

[10] Yinger, *op. cit*., cf. N. O. Brown, *Love's Body*, Random House, New York, 1966;

E. Gellner, *Contemporary Thought and Politics*, Routledge and Kegan Paul,. London, 1974; R. Wuthnow, 'Astrology and Marginality', *Journal for the Scientific Study of Religion*, Vol. 15, 1976, pp. 157–168, and *The Consciousness Reformation*, University of California Press, Berkeley, 1976; C. Hill, *The World Turned Upside Down: Radical Ideas During the English Revolution*, Penguin Books, Harmondsworth, 1975.

11 Tiryakian, 1972, *op. cit*., pp. 493, 499, 502–510; cf. E. Voegelin, *The New Science of Politics*, University of Chicago Press, Chicago, 1952, p. 133; P. Deviance, C. Fischer, E. Morin, and L. Pertossan, *La retour des astrologues*, Les Cahiers du Club Nouvel Observateur, Paris, 1971, p. 125; H. Jones, *The Gnostic Religion*, Beacon Press, Boston, 1963.

12 C. Singer, *A Short History of Scientific Ideas to 1900*, Oxford University Press, New York, 1959, pp. 389–390; E. Mendelsohn, 'The Social Construction of Scientific Knowledge', in Mendelsohn *et al*., *op. cit*., pp. 10–11; Bloor, *op. cit*., p. 113.

13 Boulding, *op. cit*.; M. Wartofsky, 'Is Science Rational?', in W. H. Truitt and T. W. G. Solomons (eds.), *Science, Technology, and Freedom*, Houghton Mifflin and Co., Boston, 1974, pp. 202–210.

14 L. Winner, *Autonomous Technology*, MIT Press, Cambridge, Mass., 1977.

15 E. Bourguignon, 'Epilogue: Some Notes on Contemporary Americans and the Irrational', in Bourguignon (ed.), *op. cit*., p. 342; cf. L. Marx, 'Reflections on the Neo-Romantic Critique of Science', *Daedalus*, Vol. 7, No. 2 (Spring), 1978, pp. 61–74; L. White, Jr., 'Science and the Sense of Self: The Medieval Background of a Modern Confrontation', *Daedalus*, Vol. 7, No. 2 (Spring), 1978, pp. 47–59; Bloor, *op. cit*., pp. 55–56; Cotgrove, *op. cit*.; Needham, 1956, *op. cit*., pp. 97–98; H. Weiman, *Religious Experience and Scientific Method*. Southern Illinois University Press, Carbondale, Ill., 1971, p. 84; Yates, *op. cit*., p. 452; cf. R. Collins, *Conflict Sociology*, New York, Academic Press, 1975b, p. 504; S. Brush, 'The Chimerical Cat: Philosophy of Quantum Mechanics in Historical Perspective', *Social Studies of Science*, Vol. 10, No. 4 (November), 1980, pp. 393–447. The volatile potential in mysticism for surrender to "higher powers" is remarked on in Veysey, *op. cit*.

16 I. Hassan, 'Culture, Indeterminacy, and Immanence: Margins of the (Post-Modern) Age', *Humanities in Society*, Vol. 1, No. 1 (Winter), 1978, p. 65.

17 J. Schumacher and R. Anderson, 'In Defense of Mystical Science', *Philosophy East and West*, Vol. 29, No. 1 (January), 1970, pp. 73–90.

18 Hassan, *op. cit*., p. 83; cf. H. Brown, *The Human Future Revisited*, W. W. Norton, New York, 1978.

19 For an example of "wholistic" sociology of science, and references to wholism in other areas of inquiry, see S. Restivo and M. Zenzen, 'A Humanistic Perspective on Science and Society', *Humanity & Society*, Vol. 2, No. 4 (November), 1978, pp. 211–236.

20 A. Cicourel, *Cognitive Sociology*, Penguin Books, Baltimore, 1973; and see Collins, 1975b, *op. cit*., p. 110; cf. A. Schaff, 'Language and Reality', *Diogenes*, No. 51 (Fall), 1965, pp. 147–167; S. Rosen, 'A Case of Non-Euclidean Visualization', *Journal of Phenomenological Psychology*, Vol. 5, 1974, pp. 33–39; S. Bochner, *Eclosion and Synthesis*, W. A. Benjamin, New York, 1969, pp. v–vi; Schilling, *op. cit*., pp. 17–18, 25–26, 32.

21 K. Jaspers, *The Future of Mankind*, University of Chicago Press, Chicago, 1963; T. de Chardin, *The Phenomenon of Man*, Harper and Row, New York, 1959, *The Divine Milieu*, Harper and Row, New York, 1960, and *The Future of Man*, Harper and Row,

New York, 1964; O. Reiser, *Cosmic Humanism and World Unity*, Gordon and Breach, New York, 1975.

[22] Forman, *op. cit.*, p. 111.

[23] Smuts, *op. cit.*; G. Hegel, *Enzyklapadie der philosophischen Wissenschaften*, Part III, Suhrkamp Verlag, Frankfurt, 1970; cf. M. Harrington, *The Twilight of Capitalism*, Simon and Schuster, New York, 1976, p. 367; Greenwood, *op. cit.*; see Cotgrove, *op. cit.*; and G. Lichtheim, 'Introduction', in G. Hegel, *The Phenomenology of Mind*, Harper Torchbooks, New York, 1967, pp. xxvi–xxvii, on wholism and totalism; G. W. Morgan, *The Human Predicament: Dissolution and Wholeness*, Brown University Press, Providence, 1968, pp. 320–330.

[24] H. Stapp, 'Theory of Reality', *Foundations of Physics*, Vol. 7, Nos. 5/6, 1977, p. 314; and see J. F. Clauser and A. Shimony, 'Bell's Theorem: Experimental Tests and Implications', *Reports on Progress in Physics*, Vol. 41, 1978, pp. 1881–1927. The reader is referred to the earlier discussion (Chapter 4) of Harvey's case study of quantum mechanics.

[25] D. Bohm, *Fragmentation and Wholeness*, Van Leer Jerusalem Foundation, Jerusalem, 1976, pp. 22–23; D. Bohm and B. J. Hiley, 'Some Remarks on Sarfatti's Proposed Connection Between Quantum Phenomena and the Volitional Activity of the Observer-Participator', *Psychoenergetic Systems*, Vol. I, 1976, p. 178; J. Sarfatti, 'Implications of Meta-Physics for Psychoenergetic Systems', *Psychoenergetic Systems*, Vol. I, 1974, p. 7.

[26] On TM, see L. Domash, *Time, Space, Causality, Geometry*, World Plan Publishing, Maharishi International University, Fairfield, Iowa, no date; on Bahá'í, see W. Hatcher, 'Science and Religion', *World Order*, Vol. 3, No. 3 (Spring), 1969, pp. 7–19, and 'The Unity of Religion and Science', *World Order*, Vol. 9, No. 3 (Spring), 1975, pp. 22–38.

[27] J. Levy-Leblond, 'Ideology of/in Contemporary Physics', H. and S. Rose (eds.), *The Radicalization of Science*, Macmillan, London, 1976, pp. 136–175; Wartofsky, *op. cit.*

[28] Harrington, *op. cit.*, L. Stavrianos, *The Promise of the Coming Dark Ages*, W. H. Freeman, San Francisco, 1976.

[29] This appendix is an abridged and edited version of my review essay on David Bohm's book, *Fragmentation and Wholeness, op. cit.*, and his two-part paper on the new order in physics published in 1971 and 1973, *op. cit.*: S. Restivo, 'The Dialectics of Physics, Knowledge, and Life', *International Society for the Sociology of Knowledge Newsletter*, Vol. 6, No. 1 (July), 1980, pp. 7–10. The quotations from Bohm are all from the material under review with one exception: the reference to his idea of reality as an "infinity of things in becoming" is from D. Bohm, *Chance and Causality in Modern Physics*, University of Pennsylvania Press, Philadelphia, 1971, p. 164 (orig. publ. 1957). The quotation from Feyerabend on Protagorean relativism is from P. Feyerabend, *Science in a Free Society*, NLB, London, 1978a, p. 28; other references to Feyerabend are from this book or his *Against Method*, Verso, London, 1978b. On Merlyn, see T. H. White, *The Book of Merlyn*, University of Texas Press, Austin, 1977, p. 125.

CHAPTER 7

[1] This chapter is partly based on S. Restivo, "The Epistemological Relevance of the Sociology of Science", prepared for the International Colloquy on Epistemologically

Relevant Internalist Sociology of Science, held at Lake Cazenovia, New York in June 1981, and organized by D. T. Campbell and A. Rosenberg, I draw directly on Hooker, 1975, *op. cit.*, for many of the ideas and terms I use in connection with this sketch of an emancipatory epistemic strategy. In Restivo and Zenzen, *op. cit.*, Hooker's work was applied to the case of sociologies of science, and his ideas modified and expanded; I introduce additional modifications in this latest attempt to expand on Hooker's work. Some of the material in this chapter has appeared in: D. Chubin and S. Restivo, 'The Mooting of Science Studies: Research Programs and Policy Analysis', pp. 53–83 in M. Mulkay and K. Knorr-Cetina (eds.), *Science Observed*, Sage, London, 1983; Restivo, *op. cit.*, 1975, and S. Restivo, 'Multiple Realities, Scientific Objectivity, and the Sociology of Knowledge', *Reflections*, Vol. I, No. 1 (Summer), 1980, pp. 61–76.

[2] M. Scheler, 'The Sociology of Knowledge: Formal Problems', in J. Curtis and J. Petras (eds.), *The Sociology of Knowledge: A Reader*, Praeger, New York, 1970, p. 177.

[3] Hooker, 1975, *op. cit.*, p. 197.

[4] This discussion follows D. Bloor, 'Durkheim and Mauss Revisited: Classification and Sociology of Knowledge', *Studies of History and Philosophy of Science*, forthcoming; quotations are from Bloor's fifty-three page draft manuscript dated September 1979.

[5] Douglas, 1970, *op. cit.*

[6] Latour and Woolgar, *op. cit.*, p. 237.

[7] Knorr, 1981, *op. cit.*, p. 17.

[8] This follows Knorr-Cetina, 1981, *op. cit.*

[9] This section is based on K. Knorr, 'Tinkering Toward Success', *Theory and Society*, Vol. 8, 1979, pp. 347–376, and 'Producing and Reproducing Knowledge: Descriptive or Constructive'?, *Social Science Information*, Vol. 16, No. 6, 1977, pp. 669–696. I have adopted Knorr's terminology for "science" and applied it to "epistemic strategies".

[10] F. Jacob, 'Evolution and Tinkering', *Science*, Vol. 196 (1977), pp. 1161–1166.

[11] Feyerabend, 1978a, 1978b, *op. cit.*

[12] K. Marx, *Economic and Philosophic Manuscripts of 1844*, Foreign Languages Publishing House, Moscow, 1956, pp. 110–111, and *Grundrisse*, Vintage Press, New York, 1973, pp. 699ff. This section is based on S. Restivo and M. Zenzen, 'Holonomy in Physics and Society', *Man-Environment Systems*, Vol. 11, Nos. 5/6, 1982, pp. 177–183.

[13] Again, I follow Hooker, 1975, *op. cit.*, closely; quotations are from Hooker unless otherwise noted.

[14] B. Barnes, *Scientific Knowledge and Sociological Theory*, Routledge and Kegan Paul, London, 1974, p. 41.

[15] Hooker, *op. cit.*, 1975, pp. 222–223.

[16] Latour and Woolgar, *op. cit.*, p. 180; Knorr, 1979, *op. cit.*, p. 369.

[17] Spengler, 1926, *op. cit.*, p. 60.

[18] C. Wright, *Wittgenstein on the Foundations of Mathematics*, Harvard University Press, Cambridge, Mass., pp. 453ff.

[19] Hooker, 1979, *op. cit.*, p. 228.

[20] Hooker, 1975, *op. cit.*

[21] Feyerabend, 1978b, *op. cit.*, pp. 82ff. Rorty argues that the distinction between necessary and contingent truths is rooted in the intellectual games of philosophers from Plato to Kant. Traditional epistemology is thus an outcome of these games and an

activity that we can readily cease indulging in. From this perspective, the relativism-realism dilemma is resolved by denying the distinction between necessary and contingent truths. Rorty carries the philosophical arguments of Quine, Sellars, and Wittgenstein toward constructivist sociology by claiming that truths are developed, sustained, and given meaning solely through social interaction and in particular conversationally: R. Rorty, *Philosophy and the Mirror of Nature*, Princeton University Press, Princeton, 1979.

[22] M. Hesse, *Revolutions and Reconstructions in the Philosophy of Science*, Indiana University Press, Bloomington, 1980, p. vii.

[23] On (a) see S. Maran, 'Pluto's Moon', *Natural History*, Vol. 88, No. 3 (March), 1979, pp. 100ff.; on (b) and (c) see F. Cajori, *A History of Mathematics*, 2nd ed., Macmillan, New York, 1919, pp. 191, 465f.; on (d) see Bloor, *op. cit.*, 1976, p. 113.

[24] G. DeGre, *Science as a Social Institution*, Random House, New York, 1955 made this distinction.

[25] R. K. Merton, *The Sociology of Science*, University of Chicago Press, Chicago, 1973, p. 269.

[26] R. LaPiere, *Social Change*, McGraw-Hill, New York, 1965, pp. 410–413, 437.

[27] C. Jencks and D. Reisman, *The Academic Revolution*, Doubleday-Anchor, New York, 1967, pp. 201–202.

[28] A. M. Carr-Saunders, *Professions: Their Organization and Place in Society*, The Clarendon Press, Oxford, 1928, pp. 30–31; E. Friedson, *Profession of Medicine*, Dodd, Mead and Co., New York, 1970, pp. 370f.; I. L. Horowitz, 'Mainliners and Marginals: The Human Shape of Sociological Theory', in L. T. Reynolds and J. M. Reynolds (eds.), *The Sociology of Sociology*, David McKay, New York, 1970, p. 345.

[29] T. Parsons, *Essays in Sociological Theory*, Rev. ed., The Free Press, New York, 1954, pp. 34–49; R. G. Francis and R. C. Stone, *Service and Procedure in Bureaucracy: A Case Study*, University of Minneapolis Press, Minneapolis, 1956, pp. 153–157; P. M. Blau and W. R. Scott, *Formal Organization: A Comparative Approach*, Chandler, San Francisco, 1962, pp. 60–63; on the convergence of bureaucratization and professionalization see, for example, W. R. Scott, 'Professionals in Bureaucracy – Areas of Conflict', in H. M. Vollmer and D. L. Mills (eds.), *Professionalization*, Prentice-Hall, Englewood Cliffs, N.J., 1966, pp. 266–267. The quotation on reliability and devotion to regulations is from R. K. Merton, *Social Theory and Social Structure*, The Free Press, N.Y., 1957, p. 200; and see Horowitz, *op. cit.*, p. 347.

[30] G. Lenski, *Human Societies*, McGraw-Hill, New York, 1970, p. 59 (and later editions); M. D. Sahlins and E. R. Service, *Evolution and Culture*, University of Michigan Press, Ann Arbor, 1960, pp. 95–97.

[31] F. Engels, *Dialectics of Nature*, International Publishers, New York, 1940.

[32] From P. Bosserman's discussion in *Dialectical Sociology*, Porter Sargent, Boston, 1968, pp. 227–237.

PART II

CHAPTER 8

[1] O. Spengler, *The Decline of the West*, Vol. I, A. A. Knopf, New York, 1926, Chapter 2; B. Hessen, 'The Social and Economic Roots of Newton's "Principia"', in N. Bukharin

et al., Science at the Crossroads, Frank Cass, London, 1931, pp. 151–212; L. Wittgenstein, *Remarks on the Foundations of Mathematics*, MIT Press, Cambridge, Mass., 1967, and *Wittgenstein's Lectures on the Foundations of Mathematics*, Harvester Press, Sussex, 1976; D. Struik, 'On the Sociology of Mathematics', *Science and Society*, Vol. 6, 1942, pp. 58–70, and 'Mathematics', in R. W. Sellars, V. J. McGill, and M. Farber (eds.), *Philosophy for the Future*, Macmillan, New York, 1949, pp. 125–152; J. Needham, 'Mathematics and Science in China and the West', *Science and Society*, Vol. 20, 1956, pp. 320–343, and *Science and Civilization in China*, Vol. III, Cambridge University Press, Cambridge, 1959.

[2] This is clearly reflected in the collected writings of "the father of the sociology of science": R. K. Merton, *The Sociology of Science*, University of Chicago Press, Chicago, 1973.

[3] K. Mannheim, *Ideology and Utopia*, Routledge and Kegan Paul, London, 1936.

[4] On radical science, see *Radical Science Journal*, published by a collective in London, *Science and Nature*, published by Lester Talkington in New York, and *Working Papers on Marxism and Science*. The "strong program" is outlined in D. Bloor, *Knowledge and Social Imagery*, Routledge and Kegan Paul, London, 1976; and see K. Knorr, 'The Ethnography of Laboratory Life: Empirical Results and Theoretical Challenges', in S. Restivo (ed.), *Newsletter of the International Society for the Sociology of Knowledge*, special issue on 'New Directions in the Sociology of Science', Vol. 7, Nos. 1 & 2 (May), 1981, pp. 4–9 for a review of ethnographic science studies.

[5] Mannheim, *op. cit.*, p. 79.

[6] W. Stark, *The Sociology of Knowledge,* Routledge and Kegan Paul, London, 1958, p. 162; Spengler, *op. cit.*, pp. 59–60.

[7] Wittgenstein, 1967, *op. cit.*, pp. 51, 80, 94, 180, 190.

[8] *Ibid.*, p. 3.

[9] *Ibid.*, pp. 4, 8; A. N. Whitehead and B. Russell, *Principia Mathematica*, 3 Vols., Cambridge University Press, New York, 1910–1913 (1st ed.), 1925–1927 (2nd ed.).

[10] Wittgenstein, 1967, *op. cit.*, pp. 20, 34, 94.

[11] *Ibid.*, pp. 45, 54.

[12] R. Hersh, 'Introducing Imre Lakatos', *The Mathematical Intelligencer*, No. 3, 1978, p. 149.

[13] I. Lakatos, 'Cauchy and the Continuum', *The Mathematical Intelligencer*, No. 3, 1978, p. 151.

[14] See, for example, E. T. Bell, *Men of Mathematics*, Simon and Schuster, New York, 1937, p. 271.

[15] Lakatos, *op. cit.*, p. 152.

[16] A. Robinson, *Non-Standard Analysis*, North-Holland, Amsterdam, 1966.

[17] Lakatos, *op. cit.*, p. 153.

[18] *Ibid.*, pp. 153–155; A. Pringsheim, 'Grundlagen der Allgemeinen Functionenlehre', in M. Burkhardt, W. Wutlinger, and R. Fricke (eds.), *Encyklopadie der Mathematischen Wissenschaften*, 2 Erste Teil, Erste Halbband, Teubner, Leipzig, 1916, p. 34.

[19] Lakatos, *op. cit.*, pp. 156–157.

[20] J. Curtis and J. Petras (eds.), *The Sociology of Knowledge: A Reader*, Praeger, New York, 1970.

[21] Stark, *op. cit.*, pp. 125, 162, 167, 194.

[22] F. Znaniecki, *The Social Role of the Man of Knowledge*, Columbia University Press, New York, 1940, pp. 190–192, 197.

[23] *Ibid.*, pp. 145–146.
[24] K. Mannheim, *Essays on the Sociology of Knowledge*, Routledge and Kegan Paul, London, 1952; and *op. cit.*, p. 43.
[25] H. J. Bos and H. Mehrtens, 'The Interactions of Mathematics and Society in History: Some Exploratory Remarks', *Historia Mathematica*, Vol. 4, 1977, pp. 10, 12–13.
[26] *Ibid.*, p. 16.
[27] R. L. Wilder, *Evolution of Mathematical Concepts*, Halsted Press, New York, 1975, pp. xvii–xix, 3; Leslie White wrote an essay in which he argued that the "objectivity" of mathematics "exists only for the individual". The locus of mathematical behavior is "the continuum of symbolic behavior", that is, "cultural tradition". Overall, this is a thoughtful essay, but it does not resolve the difficulties I discuss here. Is there an "objectivity" outside of culture, for example? White's reference to the "inevitability" of lines of inquiry in mathematics suggests that he views cultural traditions as *contexts* for the development of mathematics but not as *causative*. He hints somewhat ambiguously at the idea that mathematics manifests the social organization of mathematicians, but several remarks in his essay suggest that he was not prepared to go as far as Spengler regarding the nature of number: L. White, 'The Locus of Mathematical Reality: An Anthropological Footnote', in J. R. Newman (ed.), *The World of Mathematics*, Vol. IV, Simon and Schuster, New York, 1956, pp. 2348–2364; and see p. 2357. On mathematics as the manifestation of the social organization of mathematicians, see the suggestive remarks in S. Restivo and R. Collins, 'Mathematics and Civilization', *The Centennial Review*, Vol. XXVI, No. 3, Summer 1982, pp. 277–301.
[28] Wilder, *op. cit.*, pp. 199–201 (the ten laws); and pp. 56, 59, 99, 130, 142, 156, 171. I have added emphases and in a few cases used paraphrases here and below in my outline of Crowe's laws.
[29] This section is an expanded version of my review of this book in *Isis*, Vol. 73, No. 2, 1982, pp. 450–451; R. Wilder, *Mathematics as a Cultural System*, Pergamon Press, New York, 1981.
[30] M. Crowe, 'Ten "Laws" Concerning Patterns of Change in the History of Mathematics', *Historia Mathematica*, Vol. 2, 1975, pp. 161–166.
[31] *Ibid.*, pp. 165–166; see the twenty-three laws formulated by Wilder, 1981, *op. cit.*, pp. 127–148 in which he takes account of Crowe's laws.
[32] J. Fang and K. P. Takayama, *Sociology of Mathematics and Mathematicians*, Paideia Press, Hauppauge, N.Y., 1975, p. 32.
[33] *Ibid.*, p. 110.
[34] *Ibid.*, pp. 15–17, 118.
[35] C. S. Fisher, 'The Death of a Mathematical Theory: A Study in the Sociology of Knowledge', *Archive for History of Exact Sciences*, Vol. 3, 1966, pp. 137–138, 151.
[36] Fang and Takayama, *op. cit.*, pp. 226–238.
[37] W. Hagstrom, *The Scientific Community*, Basic Books, New York, 1965; M. Steiner, *Mathematical Knowledge*, Cornell University Press, Ithaca, 1975, p. 21.
[38] For another view, see R. Startup, 'The Sociology of Mathematics', *Sociology and Social Research*, Vol. 64, No. 2, 1980, pp. 151–167.

CHAPTER 9

[1] K. Marx and F. Engels, *The German Ideology*, International Publishers, New York, 1947, p. 35.

[2] K. Marx, *Selected Writings in Sociology and Social Philosophy*, edited by T. B. Bottomore and M. Rubel, McGraw-Hill, New York, 1956, p. 77.
[3] K. Marx, *Economic and Philosophic Manuscripts of 1844*, Foreign Language Publishing House, Moscow, 1956, pp. 110–111; *Grundrisse*, Vintage Press, New York, 1973, pp. 699ff.
[4] K. Marx, *Mathematische Manuskripte*, edited by W. Endemann, Scriptor Verlag, Kronberg Taunus, FRG, 1974; D. Struik, 'Marx and Mathematics', *Science & Society*, Vol. 12, 1948, pp. 181–196; H. C. Kennedy, 'Karl Marx and the Foundations of Differential Calculus', *Historia Mathematica*, Vol. 4, 1977, pp. 303–318.
[5] Struik, 1948, *op. cit.*, p. 190.
[6] Robinson, *op. cit.*
[7] F. Engels, *Dialectics of Nature*, International Publishers, 1940, pp. 153–154, 161–163, 198–201.
[8] F. Engels, *Anti-Dühring*, International Publishers, New York, 1939, pp. 44–46.
[9] Engels, 1940, *op. cit.*, pp. 314–315.
[10] N. Bukharin, *Historical Materialism*, International Publishers, New York, 1925, pp. 167–169; M. Cantor, *Vorlesungen uber die Geschicte der Mathematik*, B. G. Teubner, Leipzig, 1907.
[11] Spengler, *op. cit.*
[12] E. Colman, 'The Present Crisis in the Mathematical Sciences and General Outline for their Reconstruction', in N. Bukharin *et al., op. cit.*, p. 219.
[13] *Ibid.*, p. 218.
[14] *Ibid.*, pp. 222–223.
[15] *Ibid.*, p. 218.
[16] Mike B., 'Dialectics in Mathematics', *Proletariat*, Vol. 4, No. 4 (Winter), 1978, p. 36.
[17] Struik, 1949, *op. cit.*, pp. 142–143; see my discussion in Chapter 10.
[18] H. Marcuse, *Reason and Revolution*, Beacon, Boston, 1960, p. 315.
[19] C. Davis, 'Materialist Mathematics', in R. Cohen, J. Stachel, and M. Wartofsky (eds.), *For Dirk Struik*, D. Reidel Publishing Co., Boston, 1974, p. 37.
[20] *Ibid.*, pp. 40ff.
[21] C. Davis, 'Where Did Twentieth-Century Mathematics Go Wrong?', paper presented at the joint meetings of the science studies societies (HSS, PSA, SHOT, and 4S), Toronto, October 1980.

CHAPTER 10

[1] D. Struik, *A Concise History of Mathematics*, Dover, New York, 1967, pp. 18, 25; and see pp. 30–40.
[2] *Ibid.*, pp. 44–45.
[3] *Ibid.*, pp. 58–59.
[4] *Ibid.*, pp. 84–86.
[5] *Ibid.*, p. 100.
[6] *Ibid.*, pp. 104–107.
[7] *Ibid.*, pp. 117–118, 130.
[8] *Ibid.*, pp. 127–131, 137, 167.
[9] *Ibid.*, p. 139.

[10] *Ibid.*, pp. 140–142.
[11] *Ibid.*, pp. 149–180.
[12] I follow the discussion in D. Riepe, 'Dirk Struik and the Sociology of Science', in Cohen *et al., op. cit.*, p. 581.
[13] Struik, 1942, *op. cit.*, p. 58; 1949, *op. cit.*, pp. 139–140.
[14] Struik, 1942, *op. cit.*, p. 59.
[15] Riepe, *op. cit.*, p. 588; Struik, 1949, *op. cit.*, p. 138.
[16] *Ibid.*, pp. 125, 129.
[17] *Ibid.*, pp. 141, 145.
[18] Struik, 1942, *op. cit.*, p. 70.
[19] Struik, 1949, *op. cit.*, pp. 146–147.

CHAPTER 11

[1] L. Iliev, 'Mathematics in Modern Society', *Problems of the Science of Science*, Polish Academy of Science Press, Ossolineum, 1977, pp. 81–94; and see the remarks on mathematics in M. E. Omelyanovsky (ed.), *Lenin and Modern Natural Science*, Progress Publishers, Moscow, 1978.
[2] L. Hodgkin, 'Politics and Physical Sciences', *Radical Science Journal*, No. 4, 1976, p. 29.
[3] *Ibid.*, pp. 43, 48.
[4] *Ibid.*, pp. 45–46.
[5] *Ibid.*, pp. 48–49.
[6] *Ibid.*, pp. 56–58.
[7] D. Dickson, 'Science and Political Hegemony in the 17th Century', *Radical Science Journal*, No. 8, 1979, pp. 22–24.
[8] J. Needham, 'Mathematics and Science in China and the West', in B. Barnes (ed.), *Sociology of Science*, Penquin, Baltimore, 1972, pp. 22–23. See Appendix B in Part I for a critique of Needham's work.
[9] *Ibid.*, p. 24.
[10] *Ibid.*, p. 25.
[11] *Ibid.*, p. 26.
[12] *Ibid.*, pp. 27–28, 33.
[13] *Ibid.*, pp. 36, 41.
[14] *Ibid.*, pp. 43–44.
[15] M. Clagett, *Greek Science in Antiquity*, Rev. ed., Collier, New York, 1963, p. 33–34; see T. W. Africa, *Science and the State in Greece and Rome*, J. Wiley and Sons, New York, 1968, p. 20; and G. Sarton's defense of the idea is in *An Introduction to the History of Science*, Vol. I, Carnegie Institution of Washington, Baltimore, 1927, pp. 8–9.
[16] B. Farrington, *Greek Science*, Rev. ed., Pelican, Baltimore, 1961, pp. 18–19.
[17] G. Thomson, *The First Philosophers*, 2nd ed., Lawrence and Wishart, London, 1961; A. Sohn-Rethel, 'Science as Alienated Consciousness', *Radical Science Journal*, Nos. 2/3, 1975, pp. 65–101, and *Intellectual and Manual Labor*, Macmillan, London, 1978. Thomson's studies of Greek tragedy and philosophy parallel Sohn-Rethel's study of "pure reason" in Greece. Their mutual citations make it difficult at times to know who

is to be credited with what. However, Thomson's (p. 7) own testimony supports the independence and priority of Sohn-Rethel's work. Thomson does not stray far from basic Marxist premises, whereas Sohn-Rethel, for all his ingenuity and imagination, sometimes plays havoc with those premises (cf. M. Colman, 'Book review of A. Sohn-Rethel, Intellectual and Manual Labor', *Science & Society*, Vol. **43**, 1979, pp. 371–373; and C. Knee, 'Connection', a review of Sohn-Rethel's book, in *Radical Science Journal*, No. 9, 1979, pp. 100–111.

[18] K. Marx, *A Contribution to the Critique of Political Economy*, International Publishers, New York 1970.

[19] Sohn-Rethel, 1975, *op. cit.*, pp. 92–95.

[20] K. Marx, *Capital*, Vol. I, International Publishers, New York, 1967, p. 72.

[21] Thomson *op. cit.*, p. 100.

[22] Marx, 1970, *op. cit.*, pp. 28–29.

[23] *Ibid.*, p. 153.

[24] Marx, 1967, *op. cit.*, pp. 71–72.

[25] *Ibid.*, p. 82n.

[26] G. Simmel, *Philosophie des Geldes*, Duncker and Humblot, Leipzig, 1900.

[27] N. Angell, *The Story of Money*, F. A. Stokes, New York, 1929, p. 51. This critique was originally published as a review essay in Restivo (ed.), *op. cit.*, pp. 33–35.

[28] Farrington, *op. cit.*, pp. 43–45; for other examples see B. L. van der Waerden, *Science Awakening*, Science Editions, New York, 1963.

[29] G. Gorini, 'The Coin as Blazon or Talisman: Paramonetary Functions of Money', *Diogenes*, Nos. **101/102**, 1978, pp. 70–88.

[30] Angell, *op. cit.*, p. 59.

[31] Cf. A. J. Quiggen, *A Survey of Primitve Money*, Methuen, London, 1949, pp. 290–291.

[32] Marx, 1967, *op. cit.*, pp. 85, 100, 365.

[33] Sohn-Rethel, 1978, *op. cit.* p. 27.

[34] *Ibid.*, pp. 179, 201. I want to mention in concluding this a book on Marxist sociology of mathematics in Danish by Else and Jens Høyrop The draft ms of an English translation, *Mathematics in Society* (by John Lamperti, formerly of Dartmouth College and now with the American Friends Service Committee in Philadelphia), was brought to my attention by Chandler Davis (University of Toronto).

CHAPTER 12

[1] The following discussion is based on, and all quotations are drawn from, Spengler, *op. cit.*, Chapter 2. The later discussions are based on published and unpublished materials by myself and Randall Collins; see especially Restivo and Collins, *op. cit.*

CHAPTER 13 AND APPENDIX D

[1] D. Bloor, *op. cit.*, p. 108.

[2] *Ibid.*, p. 143.

[3] *Ibid.*, pp. 4–5.

[4] *Ibid*., p. 141.
[5] T. S. Kuhn, *The Structure of Scientific Revolution*, 2nd ed., University of Chicago Press, Chicago, 1970.
[6] D. Hofstadter, *Gödel, Escher, Bach*, Basic Books, New York, 1979, pp. 582–583.
[7] D. Bohm, *Fragmentation and Wholeness*, Van Leer Jerusalem Foundation: Jerusalem, 1976, p. 87.
[8] This discussion is based primarily on J. L. Richards, 'The Art and Science of British Algebra: A Study in the Perception of Mathematical Truth', *Historia Mathematica*, Vol. 7, 1980a, pp. 343–365; and 'Mathematics as a Science', 1980b, paper presented at the annual meeting of the History of Science Society, Oct. 17, 1980, Toronto, Canada.
[9] Richards, 1980a, *op. cit*. p. 346.
[10] *Ibid*., p. 360; S. Cannon, *Science in Culture: The Early Victorian Period*, Dawson Scientific History Publications, New York, 1978.
[11] This discussion is based on R, Collins and S. Restivo, 'Robber Barons and Politicians in Mathematics', *Canadian Journal of Sociology*, 1983, forthcoming; and J. W. Dauben, *Georg Cantor: His Mathematics and Philosophy of the Infinite*. Harvard University Press, Cambridge, Mass., 1979.
[12] The conjecture that mathematics and religion are intricately intertwined is nowhere more clearly expressed than in Spengler, *op. cit*., pp. 56, 66, 380, 394.
[13] See, for example, M. Kline, *Mathematics: A Cultural Approach*, AddisonWesley, Reading, Mass., 1962, pp. 582, 661, 665; cf. D. Gasking, 'Mathematics and the World', in Newman, Vol. III, *op. cit*., p. 1710.
[14] D. Bloor, 'Polyhedra and the Abominations of Leviticus', *The British Journal for the History of Science*, Vol. II, No. 39 1978, pp. 245–272; I. Lakatos, *Proofs and Refutations*, Cambridge University Press, Cambridge, 1976; M. Douglas, *Natural Symbols*, Penguin, Harmondsworth, 1973.
[15] Bloor, 1978, *op. cit*.. p. 251.
[16] *Ibid.*
[17] M. Douglas and A. Wildavsky, *Risk and Culture*. University of California Press, Berkeley, 1982, p. 138.
[18] D. Bloor, 'Durkheim and Mauss Revisited: Classification and the Sociology of Knowledge', *Studies in History and Philosophy of Science*, forthcoming (draft ms. 1979).
[19] D. MacKenzie, *Statistics in Britain, 1865–1930*, University of Edinburgh Press, Edinburgh, 1981. This is a slightly edited version of a book review that appeared in the *4S Newsletter* Vol. 7, No. 3 (Fall), 1982, pp. 48–51.
[20] Doestoevsky wrote: " ... twice-two-makes-four is not life, gentlemen. It is the beginning of death. Twice-two-makes-four is, in my humble opinion, nothing but a piece of impudence ... a farcical dressed up fellow who stands across your path with arms akimbo and spits at you. Mind you, I quite agree that twice-two-makes-four is a most excellent thing; but if we are to give everything its due, then twice-two-makes-five is sometimes a most charming little thing, too". It is not clear that this sort of exercise can be "explained" by referring to it as "merely" an expression of literary priviledge. The ability to imagine "2 + 2 = 5" does not give us "necessary truth" as would be expected if we adhered to the philosopher and mathematician William Whewell's distinction between necessary and contingent truths, proposed in 1844; F. Doestoevsky, 'Notes From the Underground', in *The Best Short Stories of Doestoevsky*, Modern Library, New York, n.d., p. 139; and see Richards, 1980a, *op. cit*., p. 362.

[21] "All representations (pictorial, verbal, realistic, abstract) are actively constructed assemblages of conventions or meaningful cultural resources that must be understood and assessed in terms of their role in activities": B. Barnes, *Interests and the Growth of Knowledge*, Routledge and Kegan Paul, London, 1977, p. 9.
[22] C. Boyer, *A History of Mathematics*, John Wiley and Sons, New York, 1968, p. 585; Struik, 1967, *op. cit.*, p. 167.
[23] A. D. Aleksandrov, 'Non-Euclidean Geometry', in A. D. Aleksandrov, A. B. Kolmogorov, and M. A. Laurent'ev (eds.), *Mathematics: Its Content, Method, and Meaning*, MIT Press, Cambridge, Mass., Vol. III, 1969, p. 183.
[24] *Ibid.*, pp. 143, 158.
[25] *Ibid.*, p. 174.

CHAPTER 14

[1] This is supported by the following historians of science and mathematics: E. Zilsel, 'The Sociological Roots of Science', *American Journal of Sociology*, Vol. 47, 1942, p. 547; D. E. Smith, *History of Mathematics*, Vol. II, Dover, New York, 1958, pp. 676–702; Farrington, 1961, *op. cit.*, pp. 153, 301; Struik, 1967, *op. cit.*, 100; M. Kline, *Mathematical Thought From Ancient to Modern Times*, Oxford University Press, New York, 1972, p. 229; and P. Boutroux, *L'ideal scientifique des mathèmaticiens*, Presses Universitaire de France, Paris, 1955 (orig. publ. 1919), p. 15.
[2] See, for example, T. B. Jones, *From the Tiber to the Tigris*, Rev. ed., Dorsey, Homewood, Ill., 1978, p. 146.
[3] Marx and Engels, *op. cit.*, p. 35.
[4] For detailed critical discussions of Marx's conception of historical materialism see: Z. A. Jordan, *The Evolution of Dialectical Materialism*, St. Martin's Press, New York, 1967; I. Meszaros, *Marx's Theory of Alienation*, Harper Torchbooks, New York, 1970; and J. McMurty, *The Structure of Marx's World View*, Princeton University Press, Princeton, N. J., 1978. For succinct accounts by Marx of his views on the core concepts of historical materialism, *mode of production, praxis*, and *praktische tatigkeit*, see Marx, 1970, *op. cit.* and 'Theses on Feuerbach', in R. C. Tucker (ed.), *The Marx-Engels Reader*, W. W. Norton, N.Y., 1972, pp. 107–109.
[5] H. Karp and S. Restivo, 'Ecological Factors in the Emergence of Modern Science', in S. Restivo and C. K. Vanderpool (eds.), *Comparative Studies in Science and Society*, Charles Merrill, Columbus, Ohio, 1974, pp. 123–143.
[6] Cf. Thomson, *op. cit.*, p. 281.
[7] Cf. Bukharin, *op. cit.*, pp. 224–225.
[8] I am indebted to many sources for the historical data on mathematics and mathematicians that I have fashioned into my narrative in the Greek and European cases, and especially the following: W. W. R. Ball, *A Short Account of the History of Mathematics*, Dover, New York, 1960; E. T. Bell, *Mathematics: Queen and Servant of Science*, McGraw-Hill, New York, 1951; Boutroux, *op. cit.*; C. B. Boyer, *The History of the Calculus and its Conceptual Development*, Dover, New York, 1949; F. Cajori, *A History of Mathematics*, 2nd ed., Macmillan, New York, 1919; T. Dantzig, *Number: The Language of Science*, 4th ed., The Free Press, 1954; J. Dedron and J. Itard, *Mathèmatiques et mathèmaticiens*, Editions Magnard, Paris, 1959; J. E. Hofmann, *The History of Mathematics to 1800*, Littlefield Adams & Co.: Totowa, N.J., 1967; Kline, 1962, 1972,

op. cit.; van der Waerden, *op. cit.*; and Wilder, 1975, *op. cit*. I have also drawn on C. C. Gillispie (editor-in-chief), *Dictionary of Scientific Biography*, Charles Scribner's Sons, New York, 1973, for some biographical data.

[9] Farrington, *op. cit*., pp. 18–19.

[10] Jones, *op. cit*., pp. 128–129.

[11] R. J. Ederer, *The Evolution of Money*, Public Affairs Press, Washington, D. C., 1964, pp. 128–129.

[12] W. McNeill, *The Rise of the West*, New American Library, New York, 1965, pp. 233–234.

[13] Wilder, 1975, *op. cit*., p. 68.

[14] Cf. Farrington, *op. cit*., p. 36.

[15] *Ibid*., p. 81; Kline, 1972, *op. cit*. p. 49.

[16] See the discussion of, for example Plimpton 322, and A06484 in van der Waerden, *op. cit*., pp. 62–81.

[17] There are problems associated with the application of ideas such as "class" and "obligarchy" in the following discussion. They are too numerous and difficult to discuss here. I follow Farrington in using such ideas by assuming that they are reasonably relevent to the social realities of ancient Greece and adequate for this exploratory analysis. Biographical information for establishing class origins and class standing in adult years is notoriously fragmented, subject to contradictory interpretations, and often entirely lacking. Nonetheless, a review of the sketches in the *Dictionary of Scientific Biography* on those persons listed in standard treatments of Greek mathematics shows that wealth and political and economic activities figure prominently in their lives. Often, enough information is available to support inference. For example, in the case of Democritus (fl. 5th century BC), it is assumed that because he is supposed to have travelled widely he must have been relatively wealthy; B. Farrington, *Science and Politics in the Ancient World*, Barnes and Noble, New York, 1966.

[18] Thomson, *op. cit*., p. 252.

[19] Farrington, 1966 *op. cit*., pp. 29–31.

[20] Struik, 1967, *op. cit*. p. 45.

[21] Farrington, 1966, *op. cit*., pp. 30–31.

[22] See, for example, R. Bonola, *Non-Euclidean Geometry*, Dover, New York, 1955 (orig. publ. 1906).

[23] Cf. Boutroux, *op. cit*., p. 53; Ball, *op. cit*., p. 53.

[24] Farrington, 1966, *op. cit*., p. 230.

[25] *Ibid*., p. 87.

[26] Farrington, 1961, *op. cit*. p. 138.

[27] Bukharin, *op. cit*., p. 184.

[28] Farrington, 1961, *op. cit*., p. 147.

[29] T. C. Heath (ed.), *The Works of Archimedes*, Dover, New York, unabridged version of the 1897 edition, n.d., pp. xv–xvi.

[30] Farrington, 1961, *op. cit*., p. 204–205; Bell, 1951, *op. cit*., p. 162.

CHAPTER 15

[1] Smith, 1958, Vol. II, *op. cit*., pp. 676–702.

[2] Farrington, 1961, *op. cit.*, pp. 204–205.
[3] *Ibid.*, p. 215.
[4] There were some notable exceptions in the Arabic-Islamic world and the East; but Byzantium civilization, for example, marked time in mathematics; Greek learning was stored but not developed. And the Hindu number system came into Europe via the Arabic-Islamic world.
[5] Farrington, 1961, *op. cit.*, pp. 214–215, 309.
[6] Cf. Marx, 1973, *op. cit.*, p. 103.
[7] Cf. Cajori, *op. cit.*, p. 271.
[8] Marx, 1967, *op. cit.*, p. 77.
[9] D. Hayes, *Europe in the Fourteenth and Fifteenth Centuries*, Holt, Rinehart, and Winston, New York, 1966; F. Braudel, *Capitalism and Material Life 1400–1800*, Harper Torchbooks, New York, 1974; P. Sweezy *et al.*, *The Transition to Capitalism*, Verso, London, 1978.
[10] J. Merrington, 'Town and Country in the Transition to Capitalism', in Sweezy *et al.*, *op. cit.*, pp. 180–181.
[11] Ederer, *op. cit.*, pp. 89–90.
[12] Struik, 1967, *op. cit.*, p. 86.
[13] Boyer, 1949, *op. cit.*
[14] Heath, *op. cit.*, xxiii.
[15] Dantzig, *op. cit.*, pp. 136–138.
[16] G. Galilei, *Le opere di Galilei Galilei*, edizione nazionale (20 vols.), Vol. 18, Barbera, Florence, 1890–1909, pp. 153–154.
[17] Ball, *op. cit.*, p. 265.
[18] Boutroux, *op. cit.*, pp. 109–111; cf. Spengler, *op. cit.*, p. 80.
[19] Braudel, *op. cit.*, p. 336; McNeill, *op. cit.*, pp. 645–647; Smith, 1958, Vol. I, *op. cit.*, pp. 437–442.
[20] Work on the calculus was by no means absent: see D. B. Wagner, 'Liu Hui and Tsu Keng-chin on the Volume of a Sphere', *Chinese Science*, Vol. III (March), 1978, pp. 59–79.

NAME INDEX

SUBJECT INDEX